BROADCAST TECHNOLOGY WORKTEXT

BROADCAST TECHNOLOGY WORKTEXT

Samuel E. Ebersole

Focal Press
Boston London

Focal Press is an imprint of Butterworth–Heinemann.

∞ Recognizing the importance of preserving what has been written, it is the policy of Butterworth–Heinemann to have the books it publishes printed on acid-free paper, and we exert our best efforts to that end.

Design, illustration, and composition by EbersoleMedia.

Library of Congress Cataloging-in-Publication Data

Ebersole, Samuel E.
    Broadcast technology worktext / by Samuel E. Ebersole.
      p. cm.
    Includes index.
    ISBN 0-240-80122-9 (pbk.)
      1. Radio. 2. Television. 3. Radio broadcasting—Equipment and supplies.
    4. Television broadcasting—Equipment and supplies.
    I. Title
    TK6550.E24 1991
    621.384'028—dc20
                                          91-23909
                                             CIP

British Library Cataloguing in Publication Data

Ebersole, Samuel E.
    Broadcast technology worktext.
    I. Title
    621.38

    ISBN 0-240-80122-9

Butterworth–Heinemann
80 Montvale Avenue
Stoneham, MA 02180

10 9 8 7 6 5 4 3 2 1

Printed in the United States of America

Contents

Contents

Preface

Where do we begin in our attempt to understand the amazing technology at work behind the scenes of a radio or television broadcast? For many of us, even those who work in this extraordinary profession, it is enough just to know that the technology does in fact work. When it doesn't, we usually know who to call to get us back on line.

This book explains the basic concepts and principles at work in the equipment used in the broadcasting profession. While at times the focus is technical, the book is in no way expected to take the place of a technical manual or to minimize the important role of those with degrees in electrical engineering. Instead, the text discusses the various equipment areas from the perspective of those who occasionally need to communicate with engineers and technicians. Knowing something about the theory behind equipment operation will endear you to those whose job it is to purchase, operate, and maintain this sometimes bewildering technology.

One reason for my writing this book is the immediate and projected shortage of qualified broadcast engineers. Although they don't always get the respect they deserve, engineers have been key players in the broadcasting industry from the beginning. They're the ones who know how and why the technology works. They're the ones we call when we can't figure out why the audio sounds thin or the talent's face looks green. Although management and production people usually don't care to know the calculus and physics at work in all of the transistors and diodes that fill our black boxes, all of us sometimes need to know the basic fundamentals at work. While engineering positions are being scaled back at many facilities and existing engineering staffs are growing older, most industry professionals agree that too few young engineers are entering the field. Some say that it is the loss of military-trained engineers. Others say that it is a failure of our educational institutions to produce properly trained graduates. Colleges and universities will continue to graduate a steady stream of broadcast journalists, but they will likely continue to shy away from technically oriented programs. For those institutions that recognize the need and attempt to meet it, I hope that this worktext will be of assistance.

For those who would like to peek inside the magician's hat, take heart—the science of broadcasting is the application of a mere handful of physical laws and theories.

Acknowledgments

The author wishes to thank former mentors and now colleagues, Harry Sova and Robert Schihl. Under their tutelage and instruction I learned to appreciate the broadcast industry and the technology on which it depends. Special thanks also to engineers Ian Hartley of KTSC-TV and John Baich of KCNC-TV for taking the time to review the manuscript and for their invaluable comments. Thanks also to Greg Hall, whose helpful suggestions were incorporated in the final manuscript. And finally, a special thanks to my wife and partner in this project, Sara Jane. Her encouragement, support and long hours at the computer transformed a stack of lecture notes into a published manuscript.

Chapter 1
TRANSMISSION

To understand the technology at work in the broadcasting industry, you must begin with the broadcast transmission. A radio or television broadcast is little more than a dispersal of radio waves over an area, an area that hopefully contains receivers that can be tuned to receive the transmission. Broadcasting, whether of radio or television signals, is a service to the community that uses a very precious natural resource. That resource is airspace, or, more accurately, allocated portions of the electromagnetic spectrum. To better understand this transmission of audio, video, and data signals via radio frequency technology, begin by looking at the radio waves themselves.

Radio Waves

Radio waves, also known as *electromagnetic waves* or *electromagnetic energy*, are made up of the component forces of electricity and magnetism, and they travel at the speed of light. Radio waves display characteristics common to both sound waves and visible light waves. In structure, a radio wave resembles a sound wave. In practice, it frequently behaves like a light wave. Some radio waves reflect off certain materials yet penetrate others. Before going too far it would be best to consider the basic properties of radio waves.

Frequency

One attribute of a radio wave is **frequency**. The frequency of a radio wave has to do with the number of oscillations, or cycles per second, and is expressed in **Hertz (Hz)**. Radio waves are found at the extreme low end of the **electromagnetic spectrum,** which includes the entire spectrum of wavelengths (from long to short) and frequencies (from low to high). The electromagnetic spectrum ranges from lowest to highest in the following order: radio, infrared, visible light, ultraviolet, X-ray, gamma ray, and cosmic ray waves. Of these, only radio waves are of interest to the broadcaster for the purpose of transmitting a signal through space. The radio wave spectrum is made up of those frequencies between 10 kHz and 300,000 MHz. The accompanying sidebar lists frequencies used for all radio communication, from military applications and amplitude modulation radio to television and communications satellite transmissions.

RADIO WAVE SPECTRUM

very low frequency
0.003 to 0.03 MHz

low frequency
0.03 to 0.3 MHz

medium frequency
0.3 to 3 MHz

high frequency
3 to 30 MHz

very high frequency (VHF)
30 to 300 MHz

ultra high frequency (UHF)
300 to 3000 MHz

super high frequency
3000 to 30,000 MHz

extremely high frequency
30,000 to 300,000 MHz

FREQUENCY: The number of oscillations or cycles per second. The frequency of a radio wave affects its physical behavior. One kilohertz is 1000 Hz; 1 megahertz is one million Hz; and 1 gigahertz is one billion Hz.

ELECTROMAGNETIC SPECTRUM: The entire spectrum of frequencies of electromagnetic waves. The electromagnetic spectrum ranges from lowest to highest in the following order: radio, infrared, visible light, ultraviolet, X-ray, gamma ray, and cosmic ray waves.

Broadcast Allocations

Within the radio wave spectrum, AM and FM radio and television transmissions are given fixed frequency and bandwidth allocations by the Federal Communications Commission (FCC). The frequency allocations for broadcast radio and television are listed for you.

Wavelength

Another attribute of radio waves is **wavelength**. Frequency and wavelength vary conversely with each other; i.e., the higher the frequency, the shorter the wavelength. Wavelength is measured as the distance from the peak of one wave to the peak of an adjacent wave. Radio waves range from several miles to a few hundredths of an inch in length. Radio waves with longer wavelengths behave much like sound waves and are subject to similar physical principles. Shorter radio waves behave much the same as light waves and exhibit many of the characteristics of infrared and visible light.

Radio waves have yet another attribute: **velocity**. The velocity of a radio wave transmitted through space is a fixed value, which is known to be the **speed of light,** or 186,300 miles per second. Because frequency, wavelength, and velocity (a fixed quantity) are interrelated, it is possible to use a simple formula to arrive at either the frequency or the wavelength of a particular radio wave when given one of the two variables. The formula is $V = W \times F$.

Velocity = Wavelength × Frequency

WAVELENGTH: A measure of distance from a point on one wave to the same point on the successive wave. In the RF realm, wavelength is determined by frequency and can be computed by using this equation: V (velocity) = F (frequency) × W (wavelength), where V is equal to the speed of light.

You can use this formula to determine the length of the radio waves generated by an FM radio station broadcasting at 99 MHz.

$V = W \times F$

V = 186,300 miles/sec or 983,664,000 feet/second
F = 99,000,000 Hertz

W = $V \div F$
W = 983,664,000 ÷ 99,000,000
$W \approx$ 9.94 feet, or 3.02 meters

Using this formula, you would find that the radio wave spectrum from 30 kHz to 300 GHz has wavelengths from 40,000 feet to 0.04 inches.

Density

The density, or magnitude of the wave is determined by the electric field and is a measure of the energy transported by the wave. The strength of the signal is determined by the number of watts (electrical power) of the radio transmitter and is a key factor in determining the clarity and reach of the broadcast transmission.

When discussing radio waves you must take into consideration the frequency (Hz, cycles per second) and wavelength (inches, feet, or meters), which are interrelated. Also consider the strength (decibels, watts), which is a variable based on the transmitter's power.

Modulation

Broadcast radio or television signals begin as an electronic signal generated in the station's studio or control room. Possible audio sources include microphones, turntables, audio tape recorders (ATRs), and compact disc (CD) players. Sources of video images include cameras and computer graphics systems. These electronic signals are combined with a radio frequency (RF) wave so they can be broadcast. The radio wave is known as the **carrier wave** (or just the **carrier**) because it carries the audio signal. The carrier wave, which has a constant frequency and intensity, is produced by an oscillator. It is important that the station maintain the exact frequency on which it has been licensed to broadcast. The process of superimposing the audio or video signal on the radio wave is known as **modulation**; this process usually takes place at the transmitter. The signal is then amplified and sent to the antenna tower. The broadcast station and the broadcast transmission engineer are held responsible for strict adherence to modulation and transmission standards.

CARRIER WAVE: The radio wave on which an audio or video signal is superimposed for broadcasting.

The two types of modulation are **amplitude modulation (AM)**, which varies the amplitude of the carrier wave while the frequency remains constant, and **frequency modulation (FM)**, which varies the frequency of the carrier wave while the amplitude remains unchanged. Both AM and FM transmissions have certain inherent advantages and disadvantages. AM transmissions tend to be more susceptible to interference from electrical storms (which cause their own radio waves of large amplitude), but on the other hand, they are as a rule capable of longer reach. AM is not as readily affected by physical obstructions as FM is, but FM has greater frequency response and a more stable signal with less static. Amplitude modulation is used for AM radio, for the picture portion of the television transmission, and for shortwave radio service. Frequency modulation is used for FM radio and the audio portion of television transmissions.

AM: Amplitude modulation. A means of superimposing an audio signal on a carrier radio wave for transmission. AM varies the amplitude of the radio wave in accordance with the signal being broadcast.

FM: Frequency modulation. A means of superimposing an audio signal on a carrier radio wave for transmission. FM varies the frequency of the radio wave in accordance with the signal being broadcast.

AM Radio

AM radio, today not nearly as popular as it once was, is a struggling contestant in the broadcasting arena. According to Radio Advertising Bureau (RAB) figures, AM listeners now make up only 23% of the overall radio listening audience, down from 75% in 1972. The reasons for this decline are complex; however, it is widely accepted that AM's technical inferiority to FM is directly responsible for much of the loss in AM's popularity. The frequency spectrum of the AM radio band ranges from 535 to 1605 kHz, with stations spaced 10 kHz apart. This makes 107 possible frequency allocations on the existing AM band. Beginning in June 1990, an AM station was required to transmit within a 20-kHz bandwidth, 10 kHz on either side of the assigned frequency. This restrictive bandwidth is partly to blame for the poor frequency response of AM broadcasts. Another and perhaps greater problem with AM fidelity is the poor quality of the AM receivers that have been standard issue for many years. Unfortunately, when the FCC was making bandwidth allocation decisions, voice transmission made up the majority of AM radio's programming day, and digital audio and CDs were a long way off. In addition to allocating specific frequencies to stations in various markets, the FCC tries to leave one unused channel between stations that might be close enough to otherwise cause interference problems. The FCC has attempted to do something to reduce the problem of interference. In the summer of 1990 a new AM band, from 1605 to 1705 kHz,

4

Carrier Signal

Constant Amplitude and Frequency

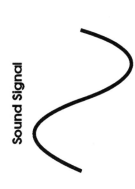

Sound Signal

Sound Signal from Audio Console

Modulated Carrier Signal

Frequency Remains Constant, but Amplitude Varies According to Sound Signal

Amplitude Modulation

Carrier Signal

Constant Amplitude and Frequency

Sound Signal

Sound Signal from Audio Console

Modulated Carrier Signal

Amplitude Remains Constant, but Frequency Varies According to Sound Signal

Frequency Modulation

Types of Modulation

(From Linda Busby and Donald Parker, *The Art and Science of Radio*. Copyright © 1984 by Allyn and Bacon. Reprinted with permission.)

went into effect, making available ten new channels on which the FCC hopes some daytime stations will find new homes. The FCC is counting on this move to free up additional space in the older band, reduce interference problems across the entire band, and provide improved signal quality.

To control interference and to regulate AM radio service, the FCC has developed classifications for AM broadcast stations and channels. There are currently four classes of stations, with several subcategories under each. They are **Class I, Class II, Class III** and **Class IV**. Class I stations (also known as **clear channel** stations) are allowed to operate with no less than 10,000 watts and no more than 50,000 watts of operating power. With this amount of power, coverage sometimes exceeds 700 miles in the evening hours. The United States has 45 clear channel allocations. Class II stations are divided into five groups: II-A, II-B, II-C, II-D, and II-S. These stations operate in the 250- to 50,000-watt range. Some Class II stations must operate at reduced power during evening hours, and others may only broadcast during daylight hours. Class III stations are for regional broadcasts and operate in the 500- to 5000-watt range. Finally, Class IV stations are intended to offer service on a local level. These stations broadcast with power in the 250- to 1000-watt range.

Class I radio stations make up a comparatively small percentage of the total number of AM stations. The majority of commercial AM stations serve small communities and operate as Class III or IV stations. Although the clear channel stations may reach very large audiences, especially at night, the smaller AM stations serve much smaller geographic areas with smaller populations.

Because there are nearly 5000 AM stations, the FCC has taken additional measures to protect them from interference from each others' transmissions. Because AM signals are sometimes carried great distances at night, nearly half of all AM stations are licensed to broadcast only during daylight hours. Also, transmitters are carefully monitored and are subject to periodic inspections to ensure compliance with power limits. Finally, directional antennas are used whenever possible. Directional antennas allow a station to direct the cast of their transmission, focusing the transmitter's energy in one or more directions—usually toward populated areas and away from other stations on the same frequency.

National Radio Systems Committee (NRSC)

As mentioned previously, AM radio has suffered over the last two decades due to its limited fidelity compared to FM. Many AM receivers have a frequency response between 2 and 3 kHz, far below FM's typical frequency response. One step toward higher fidelity for AM radio is the FCC's decision to mandate National Radio Systems Committee (NRSC) standards designed to reduce adjacent-channel interference. The committee was formed in the mid-1980s and co-sponsored by the National Association of Broadcasters (NAB) and the Electronic Industry Association (EIA), which represents consumer electronics manufacturers. By June 30, 1990, all AM stations were required to comply with the NRSC-1 standard, which reduced each station's occupied **bandwidth** from 30 to 20 kHz. This change meant that AM receivers with wider frequency response could be used with less chance of interference from the next station on the AM dial. The new wider response AM receivers will probably have frequency response bandwidths between 6 and 8 kHz. This is considered to be comparable with FM reception in an automobile

CLEAR CHANNEL: An FCC designation for a Class 1 AM station, which is permitted a maximum power of 50 kilowatts. The U.S. has 45 clear channel allocations.

BANDWIDTH: A range of frequencies within which a signal or transmission is contained. The greater the bandwidth of a transmission channel, the more information it can carry. In television, bandwidth is usually expressed in MHz.

with its high ambient noise and limited speaker fidelity. One question that comes to mind, however, is that with the rush of so many AM stations to an all-talk format, has this solution come too late? It's difficult to get excited about improved frequency response when much of the programming day is filled with telephone call-in shows.

AM Stereo

AM stereo is not a new idea, just new technology. Since the mid-1920s people have tried transmitting a stereo signal over two separate AM transmitters operating on two different frequencies. The only trouble was that it required two receivers as well, and most listeners were not ready for that level of participation, especially when FM stereo was just around the corner.

A more recent push for AM stereo has been made in response to FM's success. In the past two decades, FM has risen to dominate the music side of radio broadcasting and, in turn, much of the business as well. AM stations were left looking for a solution, and some station owners now hope stereo will be part of that solution. AM stereo actually has a few advantages over FM stereo, one of which is its longer reach. However, due to its limited bandwidth allocation, AM stereo will still offer inferior frequency response compared to FM stereo.

One reason for AM stereo's slow start is the lack of standardization. Although the FCC authorized AM stereo broadcasts back in 1982, the commission decided to take a hands-off approach to the process of selecting a standardized system. At one time, five different systems were competing for the AM stereo market, and none of them were compatible. **Motorola's C-QUAM** system has since risen to the top with over 800 stations worldwide by the spring of 1990, and that figure does not include nearly 50 more stations then in the process of converting to C-QUAM. For many stations, the conversion to C-QUAM stereo can be made for as little as $20,000. However, larger stations could incur much greater expense. According to Motorola, most car manufacturers, including Ford, General Motors, and Chrysler, are installing C-QUAM-compatible receivers in their automobiles, and 18 to 20 million C-QUAM-capable receivers are currently in place. The second most popular system, **Kahn's ISB**, was once used by nearly 100 stations. Now Kahn is reported to have only 30 stations, while it continues to battle it out with Motorola. The other systems once in contention were manufactured by Belar, Harris and Magnavox.

Another problem that is making the adoption of AM stereo a slow process is that the transmitters (stations) and receivers (listeners) must both change over to one system or another. No one wants to be an early adopter and run the risk of being left without a market or without programming. So far, the FCC has avoided stepping in to determine which way the market will go. That appears to be the approach the commission will continue to take with the ongoing policy of deregulation.

FM Radio

As stated earlier, frequency modulation is one technique used to piggyback an audio signal onto its carrier wave. This approach has several distinct advantages over amplitude modulation, not the least of which is less susceptibility to interference. The frequency spectrum for FM's allocated carrier waves ranges from 88 to 108 MHz. Incidentally, this band of frequencies falls between television channels 6 and 7. FM radio channels are assigned every 200 kHz, thus permitting a total of 100 channels. Of these, the 20 lowest channels are reserved for educational and noncommercial use. Currently, there are 3600 FM stations, but the FCC is considering making room for an additional 500 to 1000 new stations. Part of the reason for FM radio's popularity is that the bandwidth allocated to FM radio permits much greater frequency response and a greater signal-to-noise ratio than that of AM radio.

To understand the FCC's system of classifying FM radio stations, you must first understand the way that the commission has chosen to divide the United States into regions. Zone I covers the more densely populated areas of the country: primarily the Northeastern states. Zone I-A covers another densely populated area: southern California. Zone II comprises the rest of the country. Class A radio stations may operate in any of the zones, Class B stations only in Zone I, and Class C stations only in Zone II. Keeping that in mind, FM radio stations are classified according to two major criteria: effective radiated power (ERP) and antenna height. As discussed later in this chapter under "Antennas," antenna height is not an important factor for AM radio and is not a part of the AM radio classification system. However, antenna height is critical to the reach of FM broadcasts. At maximum power and antenna height, Class A FM radio stations can reach about 15 miles, Class B stations about 30 miles, and Class C stations about 60 miles. Here's a look at the various FM station classifications:

Class A stations are allowed 6000 watts maximum ERP and 100 meters maximum antenna height. The current power limit is the result of an FCC ruling that became effective in July 1989. The former limit was 3000 watts. By mid-1989, there were approximately 2000 Class A FM stations.

Class B stations are permitted 50,000 watts maximum ERP and 150 meters maximum antenna height. Remember, Class B stations are restricted to certain geographic zones, e.g., much of the Northeast region of the United States and southern California. Within the Class B category is the subcategory Class B1; these stations are permitted 25,000 watts maximum ERP and 100 meters maximum antenna height.

Class C stations may broadcast with up to 100,000 watts maximum ERP and may have a maximum antenna height of 600 meters. Within the Class C category are three subcategories: C1, C2, and C3. Their specifications are as follows:

Class C1 = 100,000 watts maximum ERP, 299 meters maximum antenna height.
Class C2 = 50,000 watts maximum ERP, 150 meters maximum antenna height.
Class C3 = New in summer of 1989, 25,000 watts maximum ERP, 100 meters maximum antenna height.

Multiplexing

Because of FM's wider bandwidth, FM radio stations can **multiplex** an independent signal along with the two channels of audio needed for a stereo broadcast. Multiplexing is a technology which allows two or more signals to be modulated on the same carrier wave. The receiver then separates the signals—in this case separating the left and right stereo signals and the auxiliary signal. A 1983 FCC ruling allows commercial and public FM radio stations to carry additional audio services on a for-profit basis. Known as subsidiary communications authorization (SCA), these services, including Muzak, electronic mail, paging, and dispatch, provide an additional source of income for FM stations. The SCA signal can only be received by users with special decoding equipment.

MULTIPLEXING: A means of transmitting two or more signals over a single wire or carrier wave.

Digital Audio Broadcasting

A new technology that has the full attention of AM and FM broadcasters is digital audio broadcasting (DAB). Although it will likely be quite a few years before DAB becomes a viable contender, its potential impact on the radio industry is great. As is the case with high-definition television (HDTV) research, the U.S. is lagging behind other countries when it comes to the deployment of this new broadcast technology. It is highly probable that Europe and Canada will begin DAB broadcasting by 1994 using the Eureka 147 system. Current projections place the first U.S. DAB stations on-line sometime in the late 1990s.

The Eureka 147 DAB system has the approval of the European Broadcasting Union (EBU) and has been successfully field tested. To reduce the transmission bandwidth required, Eureka 147 uses a source-coding algorithm for compression called *MUSICAM*. MUSICAM currently requires only 128 kilobits per second for each mono channel. (For more information on digital compression, see Chapter 9.) Eureka 147 would require European and Canadian broadcasters to look to a new frequency spectrum to begin offering the DAB service. American companies, meanwhile, are working on an alternative system that would reduce bandwidth even more. They hope to use the spectrum presently unused by FM stations or to employ multiplexing to combine analog and digital signals on the same carrier.

Television Broadcasting

Television transmission is very similar to AM and FM radio, and in fact, it shares common technology with each. The audio and video portions of the television signal are two separate signals and remain separate throughout the entire transmission process. The video signal is amplitude modulated, and the audio portion is frequency modulated. FCC approval in the spring of 1984 opened the door to the broadcasting of television audio in stereo. This technology made possible **second audio program (SAP)**, also known as **multichannel television sound (MTS)**, and allowed bilingual or narrative broadcasts. Noncommercial WTTW in Chicago became the first stereo television broadcaster in

9

August 1984. Not every station is currently equipped to produce or transmit a stereo signal, but stereo audio for television is definitely the wave of the future. Most new television receiving sets are capable of stereo reception.

The television signal requires a lot more bandwidth for transmission than radio because the video signal is much more complex than audio. AM radio transmissions fit within a bandwidth of 10 kHz, and FM radio allows 200 kHz per channel. Television, however, requires a bandwidth of 6000 kHz (6 MHz), 600 times more bandwidth than AM radio and 30 times more than FM. A portion of this 6-MHz bandwidth is reserved for the audio signal, and another small portion is reserved to guard against crosstalk, leaving 4.2 MHz for the video signal. Even with the use of modern exciters at the transmitter site, this 4.2-MHz bandwidth effectively limits the best technology to approximately 400 lines of resolution, even if the originating camera or videotape recorder (VTR) is capable of much higher resolution. And although television producers and engineers may view a high-resolution video picture on their studio reference monitors, viewers at home see far less detail. New consumer video recording formats such as S-VHS are capable of recording and playing back a slightly higher resolution signal than the one delivered over broadcast channels. If the picture quality and resolution of National Television Systems Committee (NTSC) video, once broadcast, is already surpassed by some consumer video equipment, it can't be long before consumers demand more of broadcast video. It should be no surprise then that the quest for advanced television (ATV) or high-definition television is such an important issue for broadcasters today.

Because of interference problems, one city or market cannot have stations on adjoining channels, e.g., one market would not have television channels 3 and 4 assigned to local stations. However, due to the break in the spectrum, you could have channels 4 and 5 or channels 6 and 7 in the same market. Incidentally, channel 1 is no longer allocated for television broadcasting. In 1948, channel 1 was removed from television service and assigned to land mobile, or two-way, radio service.

The **television broadcasting spectrum** is divided into two categories: very high frequencies and ultra high frequencies. Television channels 2 through 13 are VHF, and 14 through 83 compose the UHF band. The difference in frequency between the two bands creates some disparity in terms of signal strength and reach. Although UHF and VHF signals both follow a direct, line-of-sight path, the higher frequency UHF signal is more likely to be subject to interference by buildings and other obstructions. UHF is also more easily absorbed by the atmosphere and requires higher transmission power to make up for the losses. Weak signals and those subject to interference are seldom tolerated for long by viewers who have a choice. To put it another way, a viewer will almost always consider picture quality as well as program content when selecting a channel. These are just a couple of the reasons why UHF stations have historically been secondary to VHF stations in terms of audience reach, prestige, and value. Although most UHF stations continue to be priced far below VHF stations, cable television has reduced this disparity by carrying all local stations. The result has been better economic conditions for many UHF stations. And although cable companies are no longer required by law to carry all local stations, most cable operators continue to do so.

TELEVISION BROADCASTING SPECTRUM

Channel	Frequency
2 to 4	= 54 to 72 MHz
5 and 6	= 76 to 88 MHz
7 to 13	= 174 to 216 MHz
14 to 83	= 470 to 890 MHz

Other Broadcasting Services

Low-Power Television (LPTV)

Established by the FCC in February 1982, low-power television (**LPTV**) was conceived as broadcast television for special interest groups. LPTV stations were given permission to broadcast at low power on unused VHF and UHF channels. As a secondary service, LPTV stations may not interfere with full-power television stations. To ensure this, VHF LPTV stations are limited to 10 watts of power, and UHF LPTV stations to 1000 watts. This limits their effective range to 15 to 25 miles, compared to full-power stations, which can reach ten times farther. By the fall of 1990, the FCC had licensed over 930 LPTV stations. Nearly 250 of these are part of Alaska's statewide network. One LPTV station in Alaska bills itself as "The Cartoon Channel" with 24 hours of animation programming each day.

LPTV was originally intended to give minorities and small operators the chance to own a piece of the broadcast television market. However, with many LPTV licenses going to large corporations, hopes for LPTV to democratize access to the television airwaves have faded. Another unfulfilled dream was for LPTV to create a market for original local programming.

LPTV: Low-power television. These television stations were given permission to broadcast at low power on unused VHF and UHF channels. The VHF stations are limited to 10 watts of power; the UHF, to 1000 watts. This limits their effective range to 15 to 25 miles. The initial idea was to provide minorities and other special interest groups access to the broadcast media.

Instructional Television Fixed Services (ITFS)

Founded by the FCC in 1963, instructional television fixed services (**ITFS**) is a broadcasting service reserved for educational use and is located in the 2500- to 2690-MHz range. Twenty-eight channels are available to school systems for the sharing of broadcast programming. Limited interest in ITFS frequencies by educational and public broadcasters has caused the FCC to consider allocating some of these frequencies for multipoint distribution system (MDS) and other pay-television services. By November 1990, 855 ITFS facilities were licensed or permitted. One of the largest users of ITFS is the Catholic Church, which uses the signals for parochial educational services.

ITFS: Instructional television fixed services. This broadcasting service reserved for educational use is located in the 2500 to 2690 MHz range. Twenty-eight channels are available for ITFS to share broadcast programming within and among school systems.

Microwave

Higher on the RF spectrum than radio or television transmissions, microwaves are typically in the 3- to 300-GHz range. **Microwave transmissions** are frequently used in rural areas to relay audio or video signals in 30-mile (plus or minus) jumps. Before the divestiture of AT&T, almost all microwave links were owned by AT&T and were leased to broadcasters. Since deregulation, local phone companies and private industry have stepped in to fill the need for terrestrial microwave transmissions. Broadcasters frequently use microwave transmissions for their studio to transmitter link (STL). Another common use of microwave links is by remote news vans, which use microwave transmitters to send **live shots** or edited stories back to the station. **Live trucks** are frequently outfitted with microwave transmitters and have antennas mounted on a telescoping arm or pneumatic mast. This allows the operator to increase the elevation of the antenna, thus

MICROWAVES: That part of the electromagnetic spectrum lying between 300 and 300,000 MHz. Typically used by broadcasters for point-to-point, line-of-sight transmissions.

11

increasing the potential reach of the line-of-sight signal and increasing the effective working distance of the remote truck. Depending on the local topography and the elevation of the transmitting and receiving antennas, microwave shots can reach as far as 50 miles.

Propagation of Radio Waves

The wavelength of a radio wave has a great deal to do with the way that it behaves with regard to physical laws. As mentioned earlier, the lower the frequency of the wave and the longer the wavelength, the more the physical properties resemble those of sound waves. The higher the frequency and the shorter the wavelength, the more the wave behaves like a light wave. As you know from experience, sound can easily travel around and through solid objects, but light cannot. Broadcast radio waves of low frequency, commonly known as **ground waves,** can travel through the ground or through water. For example, ground waves are used by the U.S. Navy to communicate with submarines around the world. Even though low frequency radio waves pass easily through solid matter, a degree of absorption or interference occurs.

GROUND WAVES: RF transmissions of a low frequency that travel through solid objects, i.e., ground or water.

Sky waves, on the other hand, are usually of much higher frequency and do not penetrate solid materials as ground waves do. Rather than penetrate, they reflect. Sky waves that are reflected from the layer of charged particles in the upper atmosphere (ionosphere) become **skip waves**. Because of skip-wave activity, it is not uncommon to receive British Broadcasting Corporation (BBC) or other shortwave transmissions anywhere in the world. However, these transmissions are often intermittent. AM broadcasts are also subject to this phenomenon of skipping. This is one reason that many high-power AM stations are required to reduce their power at night, when skipping is more likely to occur. Because transmissions from FM stations have a higher frequency and a shorter wavelength than those from AM stations, FM broadcasts are much less likely to interfere with each other than either shortwave or AM broadcasts.

SKIP WAVES: RF transmissions, most commonly shortwave and AM broadcasts, that bounce off of the ionosphere and can be received over great distances. Skip waves occur more frequently during nighttime hours.

The highest frequency waves, known as **direct waves**, are commonly used for FM radio and VHF and UHF television transmissions. Direct waves follow a line-of-sight approach, behaving much more like light than sound. Direct waves are limited to approximately 30 miles between stations or **relay towers** (also known as *repeaters*) due to the curvature of the earth. Of course, the actual distance is dependent on the height of the tower. Microwaves are direct waves and must be relayed every 30 miles or so. Direct waves can travel thousands of miles through space, however, and are used for transmissions to and from satellites.

DIRECT WAVES: RF transmissions that are line of sight. Transmission distance is limited by curvature of the earth and by physical obstructions.

Antennas

For AM stations, the transmitting structure acts as the antenna, and its height is usually proportional to the station's frequency. Frequently the height of an AM antenna tower is one-half or one-quarter the wavelength of the station's assigned frequency. Most AM antennas are considerably shorter than 1000 feet. Although height is not an important factor for AM radio stations, just the opposite is true for FM radio and television. For

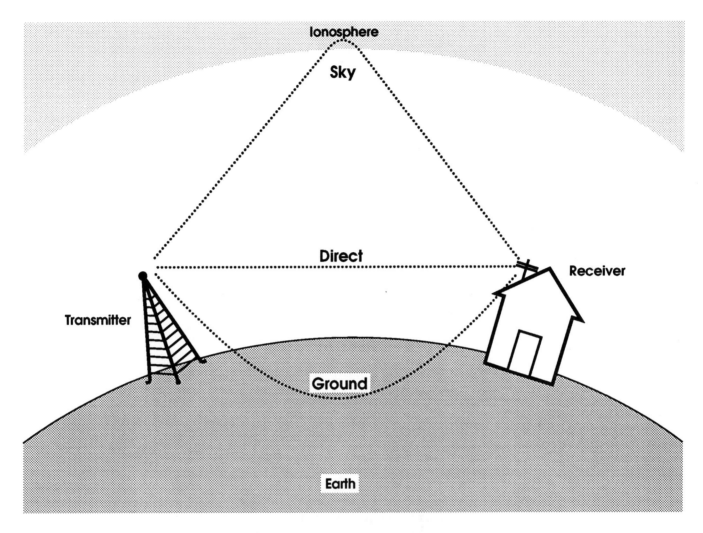

Modes of Radio Wave Propagation

(From Linda Busby and Donald Parker, *The Art and Science of Radio*. Copyright © 1984 by Allyn and Bacon. Reprinted with permission.)

these, the line of sight between the transmitting antenna and the receiving antenna is critical. Two factors that influence an FM radio or a television station's antenna's **coverage,** or **reach,** (i.e., the size of the surrounding area that is capable of receiving its broadcast) are height and directionality. To achieve height, most FM or television transmitting antennas, which are relatively short, are mounted atop towers. Others are positioned atop natural elevation peaks. For these direct wave transmissions, every 100 feet of antenna height increases its effective range by 4 miles. Incidentally, regulations on antenna height are often expressed in terms of height above average terrain (HAAT).

Broadcast antennas can be made directional much like microphones are made directional. An antenna's polar pattern may be omnidirectional, bidirectional, or unidirectional, and all to varying degrees. The intention is to direct the antenna's signal toward its intended audience and away from broadcasters who are operating on the same or adjacent frequencies. The FCC closely regulates the strength, direction, and operating hours of broadcast stations to prevent the likelihood of interference. For example, effective October 2, 1989, the FCC has ruled to increase the minimum distance separating FM stations' antennas from 105 kilometers (about 65 miles) to 115 kilometers (about 71 miles).

Another type of antenna, the **parabolic dish antenna**, is usually associated with satellite transmissions. Satellite broadcast and receiving antennas use a large, rounded dish to focus the very short waves much like a parabolic dish microphone focuses the sound waves and directs them into the microphone element. Parabolic dish antennas are extremely directional and must be positioned carefully to transmit or receive the signal. Because satellite transmissions deal with extremely long distances and high frequencies, the size of the dish and the quality of the amplification circuitry are critical to ensure satisfactory reception. Higher power satellite transmissions are now making the use of smaller dishes possible, and high-powered **direct broadcast satellites** (DBS) allow the use of receiving dishes less than one meter in diameter.

DBS: Direct broadcast satellite. Satellites powerful enough (approximately 120 watts on the Ku-band) to transmit a signal directly to a medium to small receiving dish (antenna). DBS does not require reception and distribution by an intermediate broadcasting facility, but rather transmits directly to the end user.

Satellite Transmissions

Since the launch of the Early Bird satellite (Intelsat 1) on April 6, 1965, satellites have been used for a variety of communications purposes. These uses include scientific research, weather monitoring, surveillance, and navigation. The most widespread use of satellites is for voice and data communications, including telephone communications. Broadcasters have been especially affected by satellite technology. Program distribution methods, news gathering techniques, and the relationships between networks and local stations have changed dramatically since satellites have appeared on the scene. The idea behind communications satellites is rather simple in concept, but extremely complex in execution. The satellite, once in orbit, provides a fixed point in space that can be used to reflect and relay a high-frequency radio signal. The audio, video, or data signal being transmitted actually travels out into space, is received by the satellite, amplified, and transmitted back to earth. This permits signals that can only travel in a straight line to be relayed thousands of miles to points all over the globe. Because radio waves travel at the speed of light, the thousands of miles traveled by the signal take only a fraction of a second.

BIRD: Another name for a satellite.

Communications satellites are arguably one of the most important developments ever to impact the broadcast industry, but it is not the local broadcast station or even the networks that can afford to own this technology. The manufacture, launching, and leasing of communications satellites is controlled by a few large companies, most notably GTE and Hughes Aircraft. Today's C-band **birds**, as satellites are sometimes called, cost about $125 million, while hybrid birds (those with both C-band and Ku-band transponders) go for nearly $185 million. Because satellites are an integral part of many of the new technologies, e.g., DBS and HDTV, it is likely that demand will continue to dictate price. Even with the high price of satellite broadcasts, however, there is value in choosing satellite over other forms of program distribution. Unlike other means of transmission, satellites are not limited by distance or physical or political boundaries.

Communications satellites come in either the **C-band** or **Ku-band** class. C-band satellites typically operate with a frequency of 4 to 6 GHz and fairly low power, usually only 5 to 10 watts of power per transponder. Ku-band satellites, on the other hand, operate in the 11- to 14-GHz range and use 40 to 80 watts for each transponder. It should be noted that the higher frequency Ku-band transmissions are slightly more susceptible to interference from climatic conditions than are C-band transmissions. Another difference between the two classes of satellites has to do with the operating power and dish size of each. The difference in power between C-band and Ku-band transmissions makes for a substantial difference in the size of the transmitting dish (**uplink**) and the receiving dish (**downlink**) required. C-band earth stations typically have receiving antennas 3 meters across, while Ku-band dishes can be as small as 1 meter in diameter. A facility capable of uplinking and downlinking satellite signals is commonly called a **teleport**, and international teleports are called **gateways**.

C-band technology has been available since the mid-1970s, but Ku-band satellites have only been in place since the early 1980s. For this reason, C-band users currently have many more uplink and downlink facilities at their disposal. Of the major television networks, only one, NBC, has selected Ku-band for its program distribution. However, Ku-band is catching up quickly and is expected to surpass C-band usage sometime during the middle of the 1990s. Some of the fastest growing video applications using satellite technology are satellite news gathering (SNG), business television, and educational television, and these three are heavy users of Ku-band technology. One factor that must be considered, however, is the emerging role of fiber optics for the transmission of video signals. New developments in fiber optics and compression technology could radically change the business of audio, video, and data transmission.

Most satellites have approximately 24 **transponders**, which are analogous to channels. A single transponder is capable of relaying one TV signal or more than 14,000 radio signals or about 29,000 telephone calls. Recent developments in video compression technology promise to increase the capacity of each transponder as much as tenfold. Several of each satellite's transponders are used for internal communications, supplying information necessary to keep the satellite on course and in its proper orbit.

Satellites are placed in a **geosynchronous orbit** 22,300 miles above the earth's equator. At this distance, the satellite's travel through space ensures that it remains in a fixed position relative to the earth's surface. Because the transmitted audio or video signal travels at the speed of light, one hop 22,300 miles up and back takes a little less than one-third second to complete. From its orbit in space, the satellite can relay a signal back to nearly one-third of the earth's surface, or more accurately, that part of the earth's surface covered by its **footprint**. A satellite's footprint may be large enough to cover an entire continent, or it may be much more focused, allowing for a more narrowly defined area of reception. Satellites are typically launched by rockets and, since the early 1980s, by the space shuttle program of the National Aeronautics and Space Administration (NASA). However, since the U.S. space shuttle *Challenger* was lost, NASA decided to terminate its commercial satellite launching business. The Paris-based Arianespace has taken up much of the slack, launching 22 birds between the fall of 1987 and June 1989. However, the French have not been immune to disaster. Two of their satellite-bearing rockets have blown up on the launching pad. These setbacks have caused some to predict shortages of Ku-band satellite transponders by the mid-1990s. Shortages have also driven up the price of transponders, which in the summer of 1989 were said to be going for $8 to $10 million each. This price, however, includes in-orbit and launch protection (insurance). If that

C-BAND: Typically operating at a frequency of 4 to 6 GHz, these low-power transponders operate on 5 to 10 watts of power (requiring receiving dishes about 3 meters across), although the next generation of C-band birds will have 16-watt transponders. Typically, each C-band satellite carries 24 to 36 transponders.

KU-BAND: Operating in the 11- to 14-GHz range, these are medium-power satellites, requiring about 40 to 80 watts per transponder and permitting receiving dishes as small or smaller than 1 meter across. Formerly, the demand for higher power limited the number of transponders to about ten per bird; however, the newer Ku birds have as many as 24 transponders.

UPLINK: A sending dish. A transmitter sends its signal to a large parabolic dish antenna that is aimed at the intended relay satellite.

DOWNLINK: A receiving dish. This could be a passive receiving antenna for a single household, in the case of DBS, or the antenna for the head-end of a cable system.

TELEPORT: A satellite uplink/downlink facility. Access is made available to satellite users, usually on an hourly basis. Gateways are teleports used for international satellite transmissions.

TRANSPONDER: A channel on a satellite that accepts a video or audio signal from an uplink, amplifies it, and transmits it back to earth. Most communications satellites have 24 transponders.

GEOSYNCHRONOUS ORBIT: Satellites are placed in orbit 22,300 miles over the equator and remain in a fixed position above a specific location on the globe. The satellite remains in this fixed orbit, with a few minor adjustments from time to time, until its lifetime is expired (currently about 10 years but soon to be increased to 12 years with the new generation of birds).

FOOTPRINT: The geographic area covered by a satellite's transmission signal. Most of the surface of the globe can be covered by the footprints of three satellites strategically placed.

sounds a little steep, consider that the satellite should last for 10 or 12 years. While in orbit, satellites are powered by solar cells that are designed to face the sun at all times. Small booster engines fire periodically to keep the satellite in proper position. Once the fuel is spent, there is no way to adjust the position of the satellite to keep it in its proper orbit. For this reason, the life span of a satellite is determined by the amount of fuel carried on board. By the end of 1993, several of the principal television service satellites, which were launched during the early 1980s, will run out of fuel, thus ending their useful lives.

The international communications satellite industry is controlled by an organization called *Intelsat*. It is Intelsat's job to regulate the placement of satellites and the transfer of signals across international borders. Satellites were previously positioned at 4-degree intervals, but in 1987, Intelsat agreed to switch over to 2-degree spacing (at a distance of 22,300 miles above the surface of the earth, that's about 900 miles apart). Second and third world countries that do not currently have satellites in space are concerned that all of the available parking spaces will be taken by the developed nations. The change to 2-degree spacing may help to alleviate some of the tension. By 1989, 23 new and replacement birds were scheduled for launch during the early 1990s.

Uses for Satellite Technology

Cable Television

When Home Box Office (HBO) decided to begin delivering movies to cable head-ends via satellites, it was a major breakthrough for the satellite industry. Western Union's Weststar I was the satellite of choice for HBO in 1976 when it began its enterprising move. This paved the way for numerous satellite cable networks (e.g., ESPN, USA, and CBN) and for superstations like WTBS and WWOR. Cable television is currently the largest user of satellite time for television signals.

Direct Broadcast Satellite

The idea of broadcasting a signal directly from a satellite to a small receiving dish on individual homes has been tantalizing broadcasters for some time. To make the concept feasible, however, required a satellite with a powerful enough signal that the size and price of the receiving dish could be small enough that people would be willing to make the investment. True high-powered (120 to 150 watt) DBS will require Ku-band birds more than ten times as powerful as today's C-band satellites. However, this increased power would allow receiving dishes one tenth the size of current C-band **television receive-only dishes** (TVROs). Although it has been approved by the FCC since 1982, and is currently in service in Japan, DBS in this country has gotten off to a slow start. Several DBS services are now attempting to get off the ground, including services by United States Satellite Broadcasting (USSB) and Hughes Communications. USSB hopes to have their satellite in the air by June of 1994 and service beginning the following month. For the use of these and other DBS transmissions, the FCC has allocated the

TVRO: Television receive only. Dishes that are used to receive but not transmit. Consumer TVRO dishes cost more than $10,000 in 1980, but now they can be purchased for less than $1000.

frequencies from 12.2 to 12.7 GHz. The DBS receiving dishes will be about 15 to 24 inches and, priced at $700 or less, will be inexpensive enough that the average U.S. citizen will be able to afford one.

A pseudo-DBS system is already in effect with an estimated several million dish owners receiving direct broadcasts from C- and Ku-band satellites currently in operation. You've no doubt seen these large dishes in backyards, especially in rural areas. However, this is old technology and should not be confused with what broadcasters have in mind for true DBS service.

Television Program Distribution

The use of satellites by local broadcast TV stations has grown considerably in recent years. Today, nearly 99% of all U.S. TV stations can receive C-band transmissions, and more than 70% have the ability to receive both C-band and Ku-band. The reason for this growth is that today the majority of program distribution takes place via satellite. Many syndicated programs were formerly bicycled from one station to another, but satellite technology has changed much of that. Now, many of the same programs are simply uplinked by the supplier and the local stations downlink them for live broadcast or to be taped for later broadcast. The first satellite program distribution system was established by the Public Broadcasting Service (PBS) in 1978. Network-affiliated stations receive their program feed from the network via satellite. At the discretion of the local station management, some programs are tape recorded for delayed playback while others are aired immediately as they come off of the satellite. To compensate for the different time zones across the U.S. continent, some networks provide separate feeds for each time zone. Many local stations have a manager to oversee the scheduling and technical operation of all in-coming feeds and satellite services. To manage a satellite operation successfully requires a knowledge of satellite technology and extensive contacts with satellite transmission services.

Radio Distribution

Many syndicated radio shows are now delivered by satellite rather than through telephone company land lines (**TELCOs**). In many cases, the cost is actually less than what the telephone companies were charging before the growth of satellite delivery. Some radio stations are little more than a local rebroadcast of a nationally syndicated satellite radio service; they might insert local spots and a little news now and then to give the appearance of local origination.

TELCO: Slang for the land lines leased from the telephone companies. Before the breakup of the Bell system, TELCOS were a much more common part of a local station's or a network's transmission process.

Satellite Master Antenna Television (SMATV)

Satellite master antenna television (**SMATV**) is a satellite broadcast service made available to hotels, motels, apartment complexes, condominiums, and so on. The owner pays a fee to take a feed directly from a communications satellite and then distributes the signal to the occupants or tenants. A one-dish operation can typically receive three or four cable channels. Multiple dishes or multiple feed horns allow more satellites to be

SMATV: Satellite master antenna television. A satellite broadcast service available to hotels, motels, apartment complexes, condominiums, etc.

accessed, thereby making more channels available. By 1988, nearly one million people subscribed to SMATV, and almost another million made use of the service offered by nearly 3000 hotels.

Satellite News Gathering

SNG: Satellite news gathering. A recent technology that supersedes ENG. SNG makes it possible to send live feeds from virtually anywhere in the world back to a local or network news bureau.

Satellite news gathering (SNG) has become increasingly popular in recent years for several reasons. The principal reason is readily apparent when you consider the limitations of the microwave transmitters typically used on remote news trucks. Unlike electronic news gathering (ENG) vehicles using microwave transmissions, SNG trucks are exempt from limitations of distance or topography. The reason for this is that portable SNG uplinks bounce their signal off of a satellite 22,300 miles out in space. That one hop can cover a lot of terrain and clear even the tallest skyscraper. Stations with SNG trucks can broadcast live from nearly any location to which their vehicle can be driven. The newest SNG uplink transmitters and dishes are so compact that they can be packed and taken aboard a plane like luggage. A reporter, technician, and a portable SNG uplink can broadcast from virtually anywhere on earth.

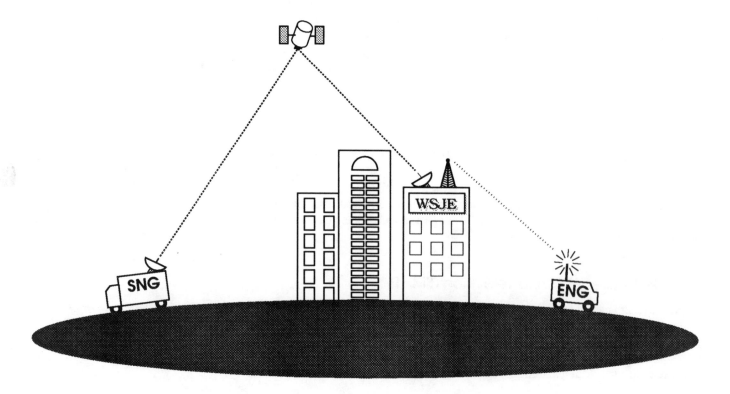

Satellite News Gathering vs. Microwave Technology

Teleconferencing

Also known as **videoconferencing**, teleconferencing is the use of live broadcast transmissions to communicate or confer by means of audio and/or video communication. Teleconferencing, in which corporate video users have invested heavily, makes use of microwave, TELCO, and satellite technology to transmit voice and picture between two or more sites, usually quite distant from each other. The idea is to save the money that would otherwise be spent on transportation, lodging, and food for a group of executives to meet in one location. Two-way teleconferencing allows complete interaction between participants. Options for videoconferencing range from simple to very complex transmission configurations. Some of the options are audio only, audio and slow-scan video, one-way video with two-way audio, and two-way video with two-way audio. The audio-only option is little more than a conference call and is quite familiar to anyone who does business by telephone. With the addition of slow-scan video, the potential of visual communication begins to be realized. Also known as *freeze-frame* or *variable-frame* video, slow-scan video allows the transmission of drawings, graphics, or any image that can be captured by a video camera. (If flat graphics are the only requirement, a conference call and fax machines could provide the same service.) With the addition of full-motion video, the full potential of videoconferencing can be seen. One-way video with two-way audio is a cost-effective approach popular with large groups. It provides audio interaction from any receiving site by means of normal telephone lines. By adding two-way video, the full interaction capability of videoconferencing is achieved but at considerable expense for the additional transmission capacity. New techniques for compression (reducing the required bandwidth) of the video and audio signals during transmission may reduce the cost of transmission considerably. Although compressed video is sometimes less than broadcast quality, it is usually quite adequate for videoconferencing.

TELECONFERENCING: Also known as videoconferencing, this is the use of audio and still or live video images as a means of conducting meetings between distant locations. The current technology uses microwave, satellite, TELCO, or fiber optic lines to convey the audio and video signals from point to point.

Fiber Optics

Fiber optics is not exactly a new technology for transmitting audio and video signals. Analog audio signals have been transmitted over fiber optic cable for quite a few years. However, it is the use of fiber optics to send digitized audio and video signals from point to point and to the home that many think may revolutionize the telephone, computer, and television industries. In fact, new technologies such as cable TV and HDTV require that broadcasters reconsider the current transmission and program delivery systems. The phone companies, long the leader in fiber-optic research and implementation, are interested in delivering much more than just telephone signals to every home in the United States. Spectrum space is limited for terrestrial broadcasting and satellite delivery systems appear to be limited by available orbiting positions in space. At the same time, broadcasters and others are looking for ways to provide more services, some of which could require a great deal of bandwidth. With more than five million miles of fiber already in place, most experts agree that fiber optics will play a large role in these developments.

The largest users of fiber optics are the phone companies. AT&T installed a transatlantic fiber optic cable that went into service late in 1988. The cable has a capacity of 300 megabits per second (Mb/s) or more than 24,000 circuits. Since 1989, coaxial cable has not been in use under the Atlantic Ocean.

It is useful to understand the technology of fiber optics, otherwise known as *optical waveguide fibers,* and its use in the transmission of analog and digital audio and video signals. Fiber optics uses photons to transmit a signal, as opposed to electrons, which are used with traditional conductive metallic cable. The photons travel from point to point through an optical waveguide, or fiber, which is composed of very pure glass. These glass fibers are thinner than human hair and weigh on the order of one ounce per kilometer.

LED: Light-emitting diode. A type of semiconductor that lights up when voltage is applied.

At either end of the transmission, the signal must be converted from or to electrical signals to interface with the current technology, which is almost entirely electronic rather than optical. Laser diodes and light-emitting diodes (**LEDs**) are used at the source, converting electronic signals into visible light. At the receiving end, photodiodes are used as detectors. This coding and decoding process allows electronic and optical transmission systems to interface with little compromise of signal integrity. Although fully digital and fully optical point-to-point transmissions may someday be a reality, the analog-electronic interface capability is still very much a part of the equation.

Fiber optic transmission has a number of advantages over traditional metallic coaxial cable. Some of these are higher capacity, minimal attenuation, less distortion, less maintenance, lower weight, smaller diameter, no ground loop problems, and less suscep-tibility to interference. Another concern about metallic cable, microwave, and satellite transmission is security. With fiber optic technology, the information content of the transmission is extremely difficult to intercept, thus making it a logical choice for military and government installations. These and other advantages of fiber optics are beginning to make it a viable choice for transmission when high performance and high information capacity are required.

With all of these advantages, you may be asking why the whole world is not cabled with fiber optics. Part of the reason is that some very difficult problems have had to be overcome, some of which have yet to see a suitable solution. One of the early problems was achieving ultrapure glass. The recent advances in fiber optic transmission have largely been due to the advances in the quality of optical glass. Corning Glass has been a leader in the field due to the incredible clarity the company has been able to achieve in its glass. To understand the clarity available in fiber optic glass, consider the amount of glass necessary to reduce the transmitted light by 10 decibels (dB). It takes just *4 inches* of window glass to cause a 10 dB loss of light. To lose the same amount of light when projecting through optical-quality glass would require a piece of glass *30 feet* thick. Compare that to the clarity of today's fiber optic glass, which experiences a similar amount of attenuation after nearly *50 miles* of glass! That's approaching a million times less light loss than that associated with ordinary window glass. Since Corning got into the fiber optic business in 1970, the company has reduced light loss from 20 dB per kilome-ter to better than 0.25 dB per kilometer.

However, ultrapure glass was just the first step in the development of a functional system. Another problem was pulse dispersion, or intermodal dispersion. This distortion of the signal results when individual light beams traveling through the fiber optic cable do not arrive at the destination at the same time. Some of the light travels right down the middle and arrives ahead of other light which bounces off the sides. These delayed signals are much like the ghosts seen in a television signal that has been reflected off of buildings or other reflective surfaces. The resulting time differences among the signals in reaching the other end is known as *pulse dispersion*. To counteract this problem, manufacturers developed a different type of optic fiber, which they refer to as *single-mode* fiber. Multimode fiber, the earlier type, is larger in diameter and allows the light to take several paths down the fiber. Single-mode fiber, on the other hand, is much thinner and allows for only one signal path. Multimode fiber is about 2/3 the diameter of a human hair, and single-mode fiber is about 1/6 the size of multimode, or approximately 0.004 millimeter. Don't let its size fool you; this tiny strand of fiber is capable of transmitting a billion digital bits per second. Compare that to the transmission rate of copper wire, which is 64,000 bits per second.

Multimode Fiber

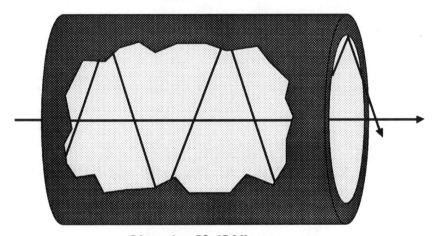

Diameter: 50-65 Microns

Single-Mode Fiber

Diameter: 9 Microns

A few difficulties still exist in plans to initiate a full-blown and expeditious implementation of fiber optic transmission. One major hurdle still to be surmounted is the cost. Fiber optic transmission is often more expensive than land, microwave, or satellite transmissions. The actual economics depend on the distance and bandwidth of the signals to be distributed. Another hurdle for fiber optics is the lack of an acceptable optical switch or splice. When you're dealing with a fiber a few thousandths of a millimeter in diameter,

21

switching and splicing optical fiber is challenging at best. Another potential problem is dirt; a microscopic speck of dirt can shut down a signal path. Telephone companies have also had to deal with the fact that the audio signal is only part of the service carried by traditional copper phone lines. Remote power, signalling, and testing are some of the others, and at present these services cannot be provided by fiber. Because of these complications, fiber transmission today is primarily a point-to-point operation, i.e., providing trunk lines between switchboards and cities but not from house to house.

Digital transmission via fiber takes place at the standard data transmission rates developed by the telephone industry. Transmission standards have been updated from DS1, also known as *T1*, which at 1.544 Mb/s provides the simultaneous transmission of up to 24 telephone conversations, to DS2 and now DS3, which operates at 44.736 Mb/s. A nonstandard DS4 is available from some manufacturers. As with any digital transmission system, it is necessary to convert the analog signal to digital pulses and back again. The device that does the conversion is known as a **codec**, which stands for coder/decoder. Codecs also perform compression, allowing reduced bit rates to be used. Currently, ABC television makes use of 45-Mb/s fiber optic cable between its Washington, D.C., and New York studios. Codecs are available that operate at 90 Mb/s, and these are preferred for extremely high-quality video transmission. Recent developments by several codec manufacturers have made it possible to compress two different video signals for transmittal over one DS3 channel. This makes two-way video feeds possible using one line.

Integrated-Services Digital Network (ISDN) and Broadband Integrated-Services Digital Network (B-ISDN)

Integrated-Services Digital Network (ISDN) is a proposed system that would provide two high-quality voice channels and one data channel per digital, fiber optic telephone line. Some time after ISDN is implemented, Broadband Integrated-Services Digital Network (B-ISDN) will offer much higher data rates, making possible multiple channels of full-bandwidth video.

Self-Study

Questions

1. Frequencies allocated by the Federal Communications Commission for AM and FM radio and VHF and UHF television broadcasting range from:
   a. 88 to 108 MHz
   b. 535 KHz to 890 MHz
   c. 108 to 535 MHz

2. When broadcasting a television signal, this method is used to modulate the audio portion of the signal on a carrier frequency:
   a. amplitude modulation
   b. frequency modulation
   c. neither AM nor FM

3. AM stereo broadcasts and receivers using this system account for most of the AM stereo broadcasting taking place today:
   a. Motorola's C-QUAM
   b. Harris
   c. Kahn's ISB

4. This type of RF transmission is limited to line of sight and is typical of microwave and television broadcasts:
   a. ground waves
   b. direct waves
   c. sky waves

5. To prevent adjacent-channel interference, which of the following TV channel combinations would not be allowed in the same broadcast market?
   a. 2 and 4
   b. 6 and 7
   c. 11 and 12

6. The effective broadcast range of LPTV stations is:
   a. 5 to 10 miles
   b. 15 to 25 miles
   c. 40 to 50 miles

7. Broadcasters using 11 to 23 GHz for terrestrial transmissions are likely using this service:
   a. microwave
   b. DBS
   c. ITFS

8. The 28 channels between 2500 and 2690 MHz available to educational and public broadcasters are known as:
   a. Public Broadcasting Service
   b. ITFS
   c. AM stereo

9. This technology allows FM radio stations to piggyback additional signals on its carrier frequency, and some of the services made possible using this technology include Muzak, electronic mail, and paging services:
   a. RF
   b. shortwave
   c. multiplexing

10. This new technology allows the program supplier to bypass any distributer—the audio and video signals are transmitted directly from the originating source to home receivers anywhere in the country using high-powered Ku-band birds:
    a. shortwave
    b. direct broadcast satellite
    c. LPTV

Answers

1. a. No. These are the frequencies allocated to FM broadcasters.
   b. Yes.
   c. No.

2. a. No. The audio signal is frequency modulated, and the video signal uses AM technology.
   b. Yes. However, the video signal is amplitude modulated.
   c. No. The audio portion of the television signal is transmitted using FM technology.

3. a. Yes. The C-QUAM system accounts for the greatest percentage of AM stereo broadcasts.
   b. No. Harris dropped out of the AM stereo race early on.
   c. No. However, Kahn's system did give C-QUAM a run for its money in the early years of AM stereo.

4. a. No. Ground waves are of a lower frequency than microwaves and can travel through solid objects.
   b. Yes. Direct waves are of higher frequency than ground waves and approach the characteristics of light waves. One such characteristic is the inability to pass through solid objects.
   c. No. Sky waves can sometimes bounce off of the ionisphere and surpass normal line-of-sight distances.

5. a. No. The spectrum allocated to channel 3 provides sufficient separation to prevent stations broadcasting on 2 and 4 from interfering with each other.
   b. No. Although channels 6 and 7 appear to be adjacent, there is spectrum space between them which is allocated to other services.
   c. Yes.

6. a. No. At 10 watts for VHF and 1000 watts for UHF, most LPTV stations can reach from 15 to 25 miles.
   b. Yes.
   c. No. The low-power status purposely limits their effective range to 15 to 25 miles.

7. a. Yes. DBS may also use these frequencies, but DBS is not a terrestrial transmission service.
   b. No. DBS is not considered a terrestrial service.
   c. No. ITFS uses frequencies from 2500 to 2690 MHz.

8. a. No. Although PBS is also known as *educational television*, it uses normal VHF and UHF frequency allocations.
   b. Yes.
   c. No. AM stereo operates on the same frequencies as AM: 535 to 1600 kHz.

9. a. No. *Radio frequency* is a term that encompasses the whole technology of broadcasting using carrier frequencies within the radio wave spectrum.
   b. No. Shortwave, 6 to 25 MHz, is a frequency for radio broadcasting that is different from either commercial AM or FM.
   c. Yes. Multiplexing allows simultaneous transmissions over sidebands.

10. a. No. Although shortwave radio broadcasts can cover great distances, only DBS technology allows audio and video signals to be transmitted directly to the receiver via satellite.
    b. Yes. Direct broadcast satellite technology uses high-powered, high-frequency transmissions to transmit a signal to a small dish mounted on the home of the program recipient.
    c. No. On the contrary, LPTV has a very limited range.

Projects

Project 1

Compute the length of the radio waves used by local AM radio, FM radio, VHF television, and UHF television stations.

Purpose

To become familiar with the relationship between broadcast frequency and the physical properties of RF transmissions.

Advice, Cautions, and Background

1. Remember the formula used to compute the wavelength: velocity = wavelength $\times$ frequency. To determine the wavelength, you must know the other two values. The velocity of radio waves is the same as the speed of light, or 186,300 miles per second, and the frequency of the station is also a known quantity.

2. You may have to call the television stations to find out their frequencies, unless you want to try to catch their sign-on or sign-off messages.

Project 2

Visit a satellite teleport facility.

Purpose

To see firsthand the facilities and operation of a satellite earth station.

Advice, Cautions, and Background

1. Depending on where you live, this project may not be feasible. If there is no teleport in your area, find out if one of the local broadcast stations has a receive-only satellite dish.

2. Find out what band the facility uses. One clue to whether they are using C-band or Ku-band is the size of their dish.

3. Ask what satellites and transponders the facility normally uses. Do they ever have to move their dish to access other satellites?

Chapter 2
AUDIO

Before you can begin to understand the specifics of audio equipment, you need to understand the nature of sound. Sound is both the beginning and the end of the audio process. Audio equipment is merely the means to capture, process, and reproduce sounds. The origin of a sound wave is a good place to begin. In most instances, sound results when energy is applied to an object. The object begins to vibrate, and these vibrations are conducted by whatever medium they encounter. Vibrations are eventually perceived as sounds when their final encounter is with the human ear. For a sound to be heard, it must first be initiated by some expenditure of energy, travel through a medium, and contact an eardrum. Some common sources of sound waves include audio loudspeakers, vocal cords, guitar strings, and tuning forks.

A sound wave is created when a moving object causes air molecules to be compressed and rarefied. Each compression is followed by a rarefaction (partial vacuum), which is followed by another compression. The distance between compressions determines the wavelength. Each cycle (from one compression to the next) sets the next in motion. A visual representation of sound waves are the ripples that result when a pebble is thrown into a pond.

In a gaseous environment, the motion of a vibrating object causes the adjacent air to begin vibrating as well. This produces changes in air pressure corresponding to (1) frequency of vibrations and (2) amplitude of those vibrations. Sound has many other characteristics (i.e., timbre, velocity, spatial quality), but they are beyond the scope of this worktext. It would be best to begin by considering the frequency of a sound wave.

Frequency

FREQUENCY: The number of oscillations or cycles per second. When dealing with acoustic sound waves, frequency is expressed in hertz (Hz) and perceived as pitch.

The **frequency** of a sound wave can be expressed as the number of times per second that the wave passes through all of its values between positive and negative. The number of cycles per second, or frequency, is typically expressed in hertz and is perceived by the human ear as the **pitch** of the sound. A frequency of 50 cycles per second (50 Hz) would be experienced by the listener as a very low tone, as opposed to 10,000 Hz, which would be experienced as a very high tone.

Concert A is 440 Hz, and middle C has a frequency of 261.63 Hz. An **octave** is equal to a doubling of pitch or frequency. Based on the assumption that the range of human hearing is between 16 and 16,000 Hz, the theoretical limits of human hearing is ten octaves. Frequencies below this lower limit are known as infrasound and frequencies above the range of human hearing are known as ultrasound. To help place the range of human hearing in perspective, the standard 88 key piano keyboard ranges from 27.5 Hz to 4186 Hz.

OCTAVE CHART		
1st Octave =	16 to	32 Hz
2nd Octave =	32 to	64 Hz
3rd Octave =	64 to	128 Hz
4th Octave =	128 to	256 Hz
5th Octave =	256 to	512 Hz
6th Octave =	512 to	1,024 Hz
7th Octave =	1,024 to	2,048 Hz
8th Octave =	2,048 to	4,096 Hz
9th Octave =	4,096 to	8,192 Hz
10th Octave =	8,192 to	16,384 Hz

You may wonder what all of this has to do with a technical understanding of professional audio equipment. The **frequency response** of specific pieces of equipment is how the manufacturer specifies the capability of the equipment to process an audio signal. Generally speaking, the microphone, amplifier, or tape machine that has the broader range of frequency response and the more linear or flatter handling across that frequency spectrum is usually the one capable of reproducing higher fidelity audio and therefore is the higher quality audio component. A frequency response curve is a graphic representation of the frequency response of a specific piece of equipment and is useful when comparing components.

Amplitude

A definition of amplitude as it relates to the audio signal would take into consideration the magnitude of the sound wave, which would be perceived by the listener as **loudness**. Loudness, then, is the subjective perception of amplitude. Generally, the greater the amplitude of the sound wave, the greater the loudness or perceived volume. A visual representation of a hypothetical sound wave might look very much like a typical sine wave, which you may recognize from math classes. As you can see by the sine wave illustration, amplitude is the height above and below the zero line, which in this case represents normal air pressure.

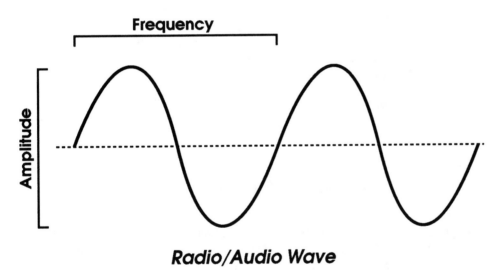

Radio/Audio Wave

Loudness is commonly measured in **decibels (dB)**. A decibel is one-tenth of a bel, a unit of measure named in honor of Alexander Graham Bell, the inventor of the telephone, among other things. However, the bel is not a very useful measure of absolute loudness. Rather, it is a subjective value used to compare the strength of two or more audio signals. To more accurately measure the amplitude of a sound wave, you must consider acoustic pressure. Acoustic pressure is measured in **sound pressure level (dB-SPL)**, which indicates the loudness above an arbitrary level of zero, the absence of sound capable of being heard by the human ear.

Remember, a decibel is not an absolute value; it is a relative measure. One decibel is equal to the smallest unit of sound level change able to be perceived by the human ear. Many consider the range of human hearing to be from 1 to 120 dB, the threshold of pain. Even though this range spans only 120 dB, it represents a tremendous dynamic range of

DECIBEL: A relative unit of measure used to compare sound levels of one signal to another. One-tenth of a Bel (in honor of Alexander Graham Bell) is a deci-Bel or dB. One dB is the smallest increase in volume that can be perceived by the human ear. An increase of 6 dB equals twice the sound pressure and is perceived as a doubling of volume.

human hearing. Because this range is logarithmic, the range of hearing is actually 1 to 10,000,000 or greater. That is to say that the loudest sound that humans can bear is nearly ten million times louder than the quietest sound they can hear. Looking at it another way, 60 dB-SPL is 1000 times louder than 1 dB-SPL because each increase of 6 dB doubles the loudness or sound pressure level.

Audio **amplification** and monitoring systems operate under a slightly different set of rules. A doubling of amperage for an amplification system does not produce a twofold increase in volume. It takes an increase of approximately ten times the power to double the perceived volume. That is one of the reasons that a small 1.5-watt amp in a car stereo can produce sounds almost as loud as those produced by a 10-watt power booster. Although other factors must be considered, such as efficiency of the loudspeakers, **distortion**, and energy loss in distribution, the fact remains that loudness is achieved at the expense of a great deal of additional power. Another factor to consider is the distance from the sound source to the listener or measuring device. Sound, like light, falls off exponentially. Sound reinforcement experts for outdoor rock concerts think nothing of using hundreds of thousands of watts of amplification.

Amplitude, or loudness, is measured and monitored by using the **volume units (VU) meter**. Some meters measure average levels, and others measure peaks. The peak-indicating meters generally have a faster response time than the 300 milliseconds of true VU meters. They have response times of 5 to 10 milliseconds, and are known as **peak program meters (PPMs)**. A very common substitute for needle-style VU meters are ones with LED readouts; one advantage is their more efficient use of space. In either case, learning to read the meters is essential to setting proper record and playback levels.

The Electronic Audio Signal

The electronic audio signal operates under the same set of physical laws as other electrical signals. A few of these laws are of special concern to audio technicians. It would be best to begin by considering impedance, signal strength, and signal-to-noise ratio. Another characteristic, whether the signal is balanced or unbalanced, is considered later in the chapter under "Audio Connectors."

Impedance

AMPLIFIER: An electronic device that increases the strength of a signal, usually the output of a preamplifier. Most amplifiers are used to drive loudspeakers or other monitoring devices. A power amplifier usually has few controls, and a control amplifier often includes integrated preamp and mixing facilities.

DISTORTION: Modification of an electronic signal that may or may not be desirable. In audio, overload distortion and total harmonic distortion (THD) are two common types of problem distortion.

IMPEDANCE: A characteristic of electrical components, rated in ohms and usually expressed as either Hi-Z (10,000 ohms and above) or Low-Z (50 to 300 ohms).

Impedance, the resistance to the flow of current in an electrical system, is measured in ohms (Ω). A low-impedance microphone has an impedance rating in the range of 50 to 600 ohms, and high-impedance microphones range from 10,000 ohms and up. The symbol Z is often used to indicate impedance—**Hi-Z** indicates high impedance, and **Low-Z** means low impedance. In almost every case, Hi-Z microphones should only be connected to Hi-Z preamplifier inputs, and Low-Z microphones to Low-Z inputs. Almost all professional broadcast-quality audio equipment is low impedance. Impedance-matching transformers can be used in situations where it is necessary to connect high-

and low-impedance equipment. The advantages of low impedance include less noise or alternating current (AC) hum from fluorescent lights or other electrical sources and longer cable runs (over 1000 feet) with minimal high-frequency **attenuation**.

ATTENUATION: To reduce the strength of a signal. A fader is a variable attenuator, and a loss pad provides a fixed amount of attenuation.

Signal Strength

An electronic audio signal is often categorized according to its strength as either a microphone- or line-level signal. Although there is plenty of room for variance within these categories, mike and line levels are somewhat standardized. The strength of a signal generated by a microphone tends to be extremely low, typically about –60 to –40 dBv (decibels referenced to 1 volt). This low output signal is known as a **microphone-level signal**. The signals generated by microphones and turntables, which also put out a very low level signal, must be sent to a **preamplifier (preamp)** before any other signal processing can take place. If you connect a microphone-level signal into a line-level input, it will not have enough signal to drive the circuit, and at best, the result will be a very weak signal with poor signal-to-noise levels.

PREAMPLIFIER: An electronic device that boosts the output of low-voltage transducers (e.g., microphones and turntable styluses) before connection to a power amplifier.

A **line-level signal** is a much stronger signal (typically 10,000 times stronger than a microphone-level signal) and is usually found at the output of magnetic tape recorder/players, mixers, or other devices with built-in preamplification. Line-level signals vary from –10 to +8 dBv or so, but they can go as high as +25 dBv. The standard for most professional, broadcast audio equipment is +4 dBv. If you accidently connect a line-level signal to a microphone-level input, you will overdrive and distort the preamp and risk damaging the components.

Many audio devices, such as **mixers** and tape recorders, allow both microphone- and line-level inputs. However, the user must make the proper selection with a switch or variable knob. If a line-level signal is patched into a microphone input, it will distort the preamp circuitry. If a microphone-level signal is patched into a line input, it will not have enough strength to provide an adequate signal.

MIXER: An electronic component that allows several audio signals to be combined. Usually contains preamps, attenuators, and tone controls for each channel so that input signals can be controlled.

In addition to microphone- and line-level signals, you should be aware of **speaker-level signals**. Speaker-level signals are those that result after a signal has been amplified by an audio power amplifier. These are intended to drive the speaker elements, which of course change the electronic signal into an acoustic signal. Speaker-level audio signals are frequently thousands, even millions of times, stronger than the original signal generated by a microphone.

Signal-to-Noise Ratio

The ratio between the signal level and the noise level of a component or sound system is referred to as the **signal-to-noise ratio**, typically abbreviated as **S/N**. The greater the signal-to-noise ratio, the greater the signal level in relation to the system noise floor. The bottom line, of course, is better audio. To simplify the concept of electronic noise and the related issue of signal-to-noise ratio, consider a situation with which you are likely to be familiar. In this analogy, the term *noise* means just that—unwanted, disruptive sounds. At a concert, a certain amount of crowd noise (coughing, whispering) is quite noticeable before the concert begins. Once the music starts, depending on the loudness of the passage, the noise becomes drowned out by the music, which represents the signal. The greater the signal-to-noise ratio, the less of a distraction the noise is to the listener. Electronic noise, like audience noise, is unavoidable. The trick is to minimize the noise level while increasing the strength of the signal in relationship to the noise.

S/N is often expressed in negative values, such as −55dB S/N. An audio component with such a rating should theoretically generate 55 dB of signal before generating 1 of dB noise. As you can see, an audio component with a rating of −55dB S/N would offer better quality audio than a similar component with a rating of −45dB S/N.

S/N is directly related to the quality of the amplification circuitry. Inexpensive circuits introduce noise into the system. Digital audio components often have S/N ratings in the 90-dB range, and many analog components have difficulty exceeding 70 dB.

Audio Sources

Audio sources for radio or television broadcasts include tape recorders (ATRs and VTRs), disk-based systems (vinyl or optical media), and of course microphones. Tape recorders, LPs, and CDs are actually audio storage devices; only the microphone has as its sole purpose the origination of an audio signal. No matter how or where a sound originates, unless it comes from a signal generator in an electronic sound synthesizer, a microphone is almost certainly the device that is used to convert the acoustic energy into electrical energy.

Microphones

MICROPHONE: A transducer that converts sound energy into electrical energy.

TRANSDUCER: A device that converts energy from one form to another. The microphone, tape head, and loudspeaker are all types of audio transducers.

DYNAMIC: A term used to describe signals that are undergoing changes in amplitude or frequency. Dynamic range has to do with the range between the extremes of amplitude or frequency of an audio signal. A dynamic microphone has a specific type of transducing element.

At its most basic level, a **microphone** is a **transducer** that converts acoustic energy or sound into electrical energy. Microphones are classified by several attributes, including transducer or element type, microphone housing, and microphone pickup pattern.

For the broadcaster, the most important transducers or element types include **dynamic** (or **moving coil**), **condenser**, and **ribbon** elements. Although all three have differing strengths and applications, the condenser microphone is probably the most popular microphone in the broadcasting business. Some of the highest quality microphones have ribbon elements, but the most simply designed and most trouble-free are the dynamic microphones.

The dynamic transducing element uses a moving coil principle. Sound waves move a diaphragm that is connected to the voice coil. As this coil of wire moves in a magnetic field, it causes very slight voltage changes in the coil. As with all types of microphone transducers, this signal is very weak and must be amplified many times before it can be processed. Because of this, the microphone preamp is usually the first electronic device that the signal encounters once it leaves the microphone. The quality of the microphone preamp can be crucial to the sound quality generated by the microphone.

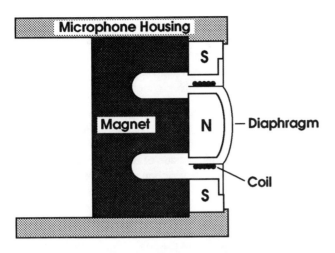

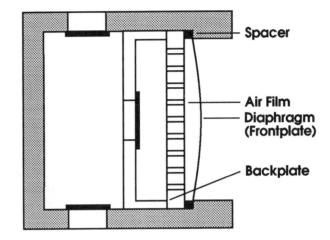

Side View of the Element of a
Moving-Coil Microphone

Cross Section of the Element of a
Capacitor Microphone

Microphone Element Types

(From *Audio in Media*, 2/E, by Stanley R. Alten © 1986 by Wadsworth, Inc. Adapted/Reprinted by permission of the publisher.)

Some of the attributes of the dynamic microphone element are the following: it is rugged and withstands rough treatment; it is reasonably priced; it has little sensitivity to handling noise; it tends to emphasize sibilance; and it is able to withstand close miking and high sound pressure levels. A classic example of a dynamic microphone is the hand-held Electro-Voice 635A, a long-time favorite of broadcast journalists.

The condenser microphone (more accurately called a capacitor microphone) has become extremely popular and is most easily distinguished from the dynamic microphone in that it requires some type of power source. The power source more often than not is provided by a small, onboard battery. The other option for supplying power to a condenser microphone is to use something called **phantom power.** Phantom power is so named because the power source is not enclosed within the microphone housing. Instead, a mixer or sound board capable of providing phantom power sends the proper DC voltage (usually between 11 and 48 volts DC) down the audio cable to the microphone, making the use of onboard batteries unnecessary. Most condenser microphones use a diaphragm and an electret backplate configuration. Between the diaphragm and the backplate lies a narrow air space. Once a polarizing voltage is applied to either the diaphragm or the backplate, any movement of the diaphragm causes a change in capacitance between the two plates. This produces a very low voltage, high-impedance output that must then be boosted by a miniature amplifier. In most cases, this tiny preamplifier is built into the microphone housing itself. *Electret* condenser microphones are condenser microphones which do not require an external polarization voltage to be applied to the backplate. Rather, a permanent electric field is created within the microphone capsule. An internal or external power source remains a requirement to power the on-board preamplifier.

Some attributes of the condenser microphone are the following: it is usually more expensive than dynamic microphones; it requires a power supply (either battery or phantom power); it generally has excellent frequency response; it is moderately rugged; and it can be contained in a very small housing. This last point explains why almost all high-quality lavaliere microphones use condenser transducing elements.

The third type of transducing element is found in the ribbon microphone. The name comes from a thin metal ribbon inside the microphone that vibrates according to the air pressure of the audio source. Early microphones that used ribbon elements were very sensitive and fragile and therefore were seldom found outside of recording studios. However, recent improvements have resulted in excellent ribbon microphones that are nearly as durable as the other designs.

Some of the attributes of the ribbon microphone are the following: it is generally very sensitive and has excellent transient response; earlier models were fragile and could not stand rough treatment; it is typically used in the controlled environment of the studio; and it is sometimes used for individuals who have popping problems with certain consonants such as *P*'s and *T*'s. The classic example of a ribbon microphone is the RCA 77DX microphone of early radio fame.

One additional type of microphone transducing element is the boundary or **pressure zone microphone,** commonly known as a **PZM®** (trademark of Crown Ltd.). It is different from the other microphones in that it is designed to pick up sound vibrations through solid objects. PZM microphones are frequently attached to Plexiglas® sheets and used for miking large areas (e.g., choirs). They are sometimes placed on table tops to pick up people seated around the table.

Polar Patterns

All microphone elements are essentially nondirectional, that is, they do not favor sound coming from any one direction but rather pick up sounds equally well from every direction. (One exception to that rule is the ribbon microphone, which by nature has a bidirectional pickup pattern.) Regardless of whether you choose to use a dynamic, condenser, or ribbon microphone, it is sometimes desirable to have a microphone that can be positioned in such a way as to hear certain sounds but reject sounds coming from other directions. Except in the case of the ribbon microphone, it is not the transducer that determines the directionality or selectivity of the microphone but rather the microphone housing. The directionality of a particular microphone is often expressed by its **polar pattern**—a two-

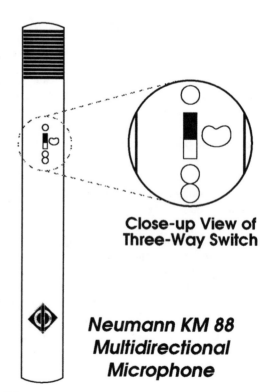

Close-up View of Three-Way Switch

Neumann KM 88 Multidirectional Microphone

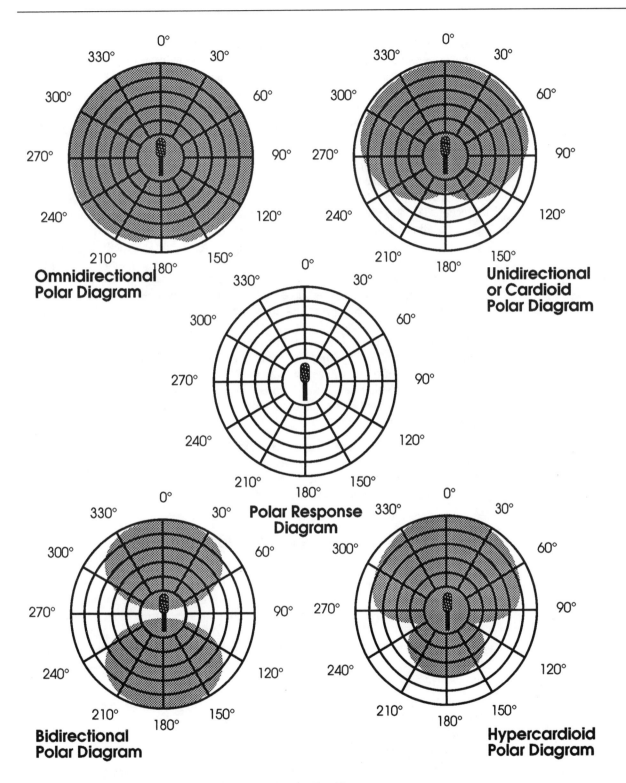

Omnidirectional Polar Diagram

Unidirectional or Cardioid Polar Diagram

Polar Response Diagram

Bidirectional Polar Diagram

Hypercardioid Polar Diagram

Pickup Patterns

(From *Audio in Media*, 2/E, by Stanley R. Alten © 1986 by Wadsworth, Inc. Adapted/Reprinted by permission of the publisher.)

OMNIDIRECTIONAL: A microphone pickup pattern that is equally sensitive to sounds coming from all directions.

UNIDIRECTIONAL: A microphone pickup pattern that is sensitive to sounds coming from one primary direction.

CARDIOID: A unidirectional pickup pattern for microphones. Named for its heart-shaped pattern, which is most sensitive to sounds directly in front, less sensitive to sounds coming from the sides, and least sensitive to sounds from the rear.

dimensional diagram of the microphone's pickup pattern. The three principal polar patterns are **omnidirectional** (the prefix *omni* refers here to all or every, as in *all directions*), **bidirectional**, and **unidirectional**. Some microphones can even be modified from one pattern to another by changing the configuration of the casing or by flipping a switch on the microphone itself. As illustrated by the accompanying diagram of the Neumann multidirectional microphone, the symbols for omnidirectional, bidirectional, and unidirectional are commonly indicated by the symbols $\bigcirc$, ∞ and $\bigcirc$ respectively.

Omnidirectional microphones are equally sensitive to sound coming from any direction, bidirectional microphones are most sensitive to sound coming from two directions, and unidirectional microphones are most sensitive to sound coming from one primary direction. The degree to which the unidirectional microphone rejects sounds coming from other than the primary direction is determined by its degree of directionality. The range is quite broad, from nearly omnidirectional to highly directional. The **cardioid** polar pattern, named for its resemblance to the shape of a heart, is a fairly common unidirectional pattern. It is followed by the more directional **supercardioid** and **hypercardioid**; the latter is also commonly known as a **shotgun microphone** due to its focused pattern of sound selection. Unidirectional microphones are selected when it is desirable to increase the working distance of the microphone, that is, the range at which the microphone will still pick up acceptable quality audio. In theory, a microphone with a cardioid pickup pattern has an increased working distance of approximately 1.7 to 1. That is to say that the same quality of sound will be achieved by an omnidirectional microphone placed 10 inches from the sound source and a cardioid microphone placed 17 inches from the same source. A supercardioid pickup pattern should exhibit an increased working distance of approximately 2:1.

It is important to note that the omnidirectional microphone usually has the flattest and smoothest frequency response and is less susceptible to breath pops. For these reasons it should be used whenever possible. However, the advantages of the unidirectional microphones in eliminating unwanted sound is often a more important consideration.

Output Level and Frequency Response

The output level of a microphone is usually rated in dBm, an electrical measurement of power that corresponds to the output in decibels related to 1 milliwatt at a standard sound pressure of 10 dynes/cm^2. This is the value given for the approximate sound pressure impressed on a microphone by normal conversation 12 inches from the microphone. Microphone output is usually much lower than 1 milliwatt. In fact, the output level of most microphones is in the range of 0.001 to 0.01 volts. Generally speaking, condenser microphones often have a slightly higher output level than dynamic microphones, which have a higher output than ribbon microphones.

FREQUENCY RESPONSE: The manner in which an audio component responds to its source, i.e., whether certain frequencies are enhanced or attenuated. A microphone with a flat response responds equally to frequencies across a broad spectrum. A frequency response curve is a visual graph representing the performance of a piece of electronic equipment, e.g., a microphone.

As stated earlier, all microphones convert acoustic energy into electrical current. How accurately they do this is best determined by noting the microphone's **frequency response**. The frequency response of a microphone is usually charted on a graph with response in dB on the vertical axis and a frequency scale in Hz along the horizontal axis. A straight horizontal line indicates a flat response. Such a microphone contributes very little sound coloration to the sound entering the microphone. As illustrated by the accompanying frequency response curve, the Electro-Voice RE20 broadcast-quality microphone has a fairly flat response, i.e., all frequencies between 50 Hz and 16 kHz are

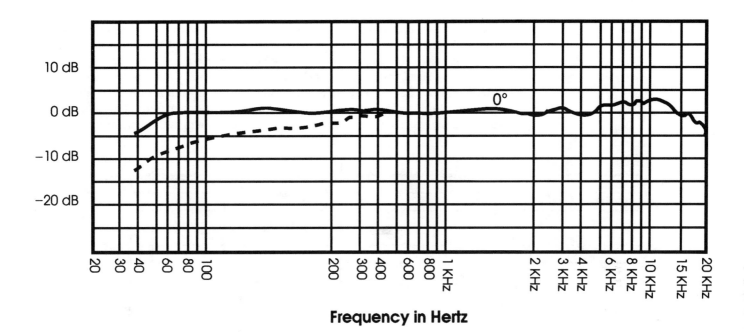

Frequency in Hertz

Frequency Response for the EV RE20 Microphone

evenly represented. Be aware, however, that some microphones intentionally emphasize or de-emphasize certain frequencies to achieve a certain effect. An example of this is demonstrated by most lavaliere microphones, which artificially boost frequencies around 4 kHz to increase speech intelligibility and to compensate for the deficiency in certain frequencies caused by the microphone's usual placement on the upper chest of the talent. However, any intentional frequency boost or cut to enhance performance must be made carefully and must allow smooth transitions if the microphone is to be of professional quality.

Frequency response is measured for audio amplifiers and other components as well. A "perfect" amplifier would have a frequency response rating of ±0 dB, 20 to 20,000 Hz. That is, there would be no variation of the output level across the entire spectrum of input frequencies. For microphones and consumer audio gear, a frequency response rating of ±3 dB over a limited range (e.g., 50 to 18,000 Hz) may be more realistic.

Proximity Effect and Bass Roll-Off

Proximity effect is an attribute associated with most unidirectional microphones—the closer the sound source to the microphone, the greater the bass response. Some listeners perceive the boost in bass frequencies as greater presence, or a feeling that the sound source is closer to the listener. This presence is considered by some announcers to be an advantage of the unidirectional microphone and is exploited for effect. Proximity effect is created by the ports on the side of the microphone, the same ports that make the microphone directional. However, when this additional bass response is undesirable, it can be defeated in certain microphones. To defeat the proximity effect, these microphones have a built-in **bass roll-off** switch, which is nothing more than a small, onboard equalizer. As seen on this Sennheiser MK2 microphone, the bass roll-off switch has three settings: position I is for flat response (no bass roll-off), position II is for a 7-dB cut at 50 Hz, and position III is for a 20-dB cut at 50 Hz. The bass roll-off switch can be useful for reducing other unwanted low frequencies, such as air conditioning rumble or wind noise.

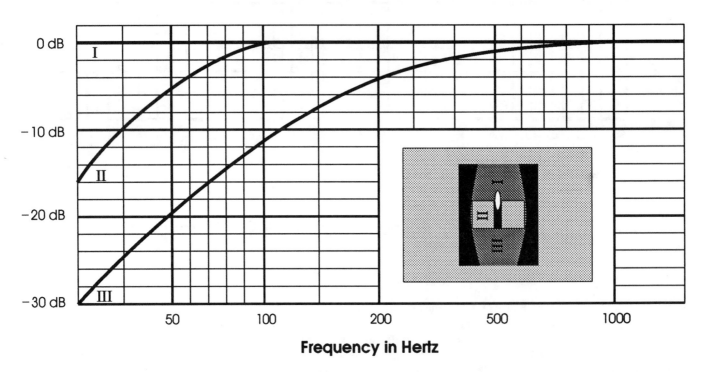

Frequency in Hertz

Bass Roll-Off Switch and Effect on Frequency Response of Sennheiser MK 2-3

(Copyright, Sennheiser. Adapted with permission.)

Frequency control or filtering is sometimes expressed in terms of what part of the signal is passed through unchanged rather than what part of the signal is boosted or cut. For example, bass roll-off is also known as *high pass*. The symbol for bass roll-off or high pass is ⌐, and the symbol for low pass or high frequency attenuation is ⌐.

Phase Cancellation

An electronic audio signal, like the sound wave that it represents, has a characteristic that is referred to as *phase* or *polarity*. The phase of the electronic signal varies with the phase of the acoustic signal. It is often graphically represented by a sine wave with the peak representing the positive polarity and the trough representing the negative polarity. It is a fairly simple matter for the phase of the electronic signal to be reversed simply by reversing the wires conducting the signal. This can take place within the microphone

38

electronics or within the cable transmitting the signal to the audio mixer. Once the positive and negative wires or contacts have been switched, the positive acoustic pressure is represented by the negative electronic signal, and vice versa. This is not necessarily a problem until you try to mix two audio signals that are generated by the same audio event but that are out of phase with each other. Mixing two signals that are of the same intensity but 180 degrees out of phase with each other will result in the two signals effectively canceling one another. The practical result is a thinness or loss of lower frequencies of the resulting mixed signal. Because electronic phase is easily corrected—many audio mixers have phase reversal switches built into each channel—the challenge here is simply to be able to hear and recognize phase cancellation when it occurs. It may also be a good idea to make sure that all of your audio cables are wired the same way.

A closely related situation is encountered when one mikes a sound source with more than one microphone. Because the phase of the acoustic audio wave varies with distance, incorrect placement of the microphones may result in acoustically out-of-phase signals being generated. When the signals are summed at the mixing board, the result can be very similar to the electronic phase cancellation mentioned earlier. To prevent this from happening, make sure that the distance between microphones is at least three times the distance of the microphone to the audio source. For example, if the first microphone is 1 foot from the sound source, the second microphone should be at least 3 feet from the first microphone. Remember, acoustic phase cancellation should not be totally eliminated— after all, it is an important part of the stereo imaging effect. However, excessive phase cancellation that results in low frequency loss should be avoided.

Turntables

Perhaps little needs to be said about turntables due to their dubious future in most broadcasting applications (except perhaps for those stations that play Golden Oldies). The demise of the turntable and vinyl media is largely due to the success of the CD and the arrival of digital technology. Professional turntables, as mentioned earlier, have a very low level signal and so must have a high-quality preamp for good audio fidelity. This preamp is almost always a separate component from the turntable itself. Many audio boards have built-in turntable preamps available on one or two channels. On the input pad for these channels, the proper preamp setting is usually indicated by a position labeled "phono."

Professional turntables are in part differentiated from consumer models in that they have a fast start-up time, necessary to avoid dead air or **wow**. Wow is a change in pitch caused by a deviation in the speed of the turntable's drive. Although most professional turntables are direct drive, rim-drive and belt-drive models are also used. Some very high quality turntables employ very sophisticated belt-drive mechanisms; the advantage is that they can be made slightly quieter than direct-drive models. Most professional turntables offer quartz-locked pitch control.

The transducing part of the turntable is the stylus and cartridge assembly. The stylus, sometimes referred to as the *needle*, is available in two basic types: spherical and elliptical. The spherical stylus is rugged and contributes less to record wear. The elliptical stylus offers higher fidelity at the price of shortened record life.

WOW: A problem with almost any mechanical recording/playback medium, notably turntables or ATRs, that is experienced as a variation in pitch due to speed fluctuations.

FLUTTER: An instability or variation of a mechanical transport, usually in a turntable or tape machine, and usually between 5 and 15 Hz. Introduces undesirable distortion to the signal.

Compact Disc

The **compact disc (CD)** is a digital, optical technology. Since its debut nearly 10 years ago, the audio CD has been commercially available as a read-only storage medium. Recently, however, several manufacturers have released models with recording capability. Some of the advantages of the optical format over conventional vinyl disks include greater resistance to physical damage, no surface noise, and of course, much higher fidelity. The CD is 4.7 inches in diameter, has 80 minutes of playing time on one side only, and spins at 500 to 200 rpm (the speed decreases as the scanner moves from the inner tracks outward). Due to the CD's indexing capability, it is gaining popularity as the source machines for automated playback configurations.

Audio Tape Recorders

Audio tape recorders (ATRs) are the workhorses for much of the audio production in broadcast operations. Available in both analog and digital, ATRs come in a wide assortment of formats, tape widths, and recording speeds. The three basic types of ATRs are reel-to-reel, cartridge, and cassette. Cartridge and cassette recorders are fairly standardized, but reel-to-reel ATRs come in a wide array of formats and serve numerous functions. (For more in-depth treatment of audio recording and recording formats, see "Audio Tape Recording" in Chapter 6.)

Signal Processing

Equalization

Several types of audio **equalization (EQ)** are available, and although they all serve to allow the user to boost or cut certain frequencies, they vary greatly in the amount of control over that process. The most basic type of equalization is the tone control. Basic car stereos often use a tone control knob to allow the user simple control of the bass and treble frequencies.

The next step up from simple tone control involves two-way and three-way EQ. Mixers, amplifiers, or other audio devices that have three-way equalization provide separate controls for high, middle and low frequencies, which can be either boosted or cut as desired.

EQUALIZER: A device that allows you to increase or attenuate certain frequencies of an audio signal. Graphic, selectively variable, and parametric are all types of equalizers. Often expressed as *EQ*.

A greater degree of control and a more visual display of the EQ process are available with the **graphic equalizer**, which can be easily distinguished by its array of sliders, each of which represents a specific frequency range. Different models of graphic equalizers are available with varying levels of control. Some divide the frequency spectrum of 20 Hz to 20 kHz into as few as six ranges, while others use fifteen or more divisions. Of course, the more divisions, the more precise the control afforded and the narrower the band of frequencies affected by each slider.

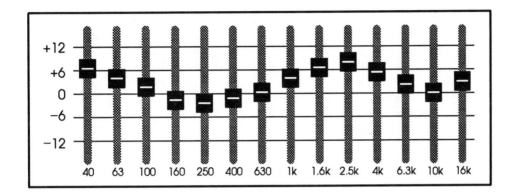

Graphic Equalizer

The **selectively variable** EQ is usually found on higher priced audio consoles. For each frequency group (high, middle, low) there are two knobs. One allows you to select the specific frequency you wish to adjust, and the other is used to specify the amount of boost or cut.

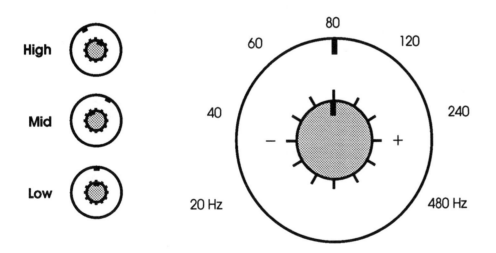

Enlarged View of Low-Frequency Control

Selectively Variable Equalizer

Finally, the most precise equalization control is offered by the **parametric** EQ. It is similar in design to the selectively variable EQ, but it has one more variable. It allows you to select the width of the bandwidth you wish to control—ranging from narrow to broad. This variable is known as the **Q factor**. As you can see in the following diagram, the parametric EQ allows you to boost or cut a very narrowly defined frequency range, acting much like a notch filter, or it may be more broadly defined for a gentler shaping of the wave characteristic.

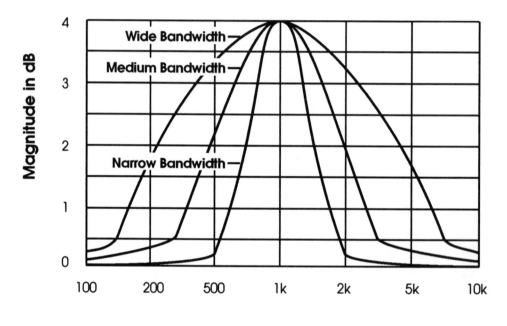

Frequency in Hz

Parametric Equalization

Another device used to shape the frequency response of an audio signal is the **notch filter**. Often used to solve a problem, the notch filter gets its name because of the way in which it cuts a "notch" out of the frequency spectrum, as shown by the frequency response graph in the diagram following. A notch filter must be used carefully, but it can be a lifesaver when a 60-Hz AC power hum or other frequency-specific noise is interfering with an otherwise clean audio signal.

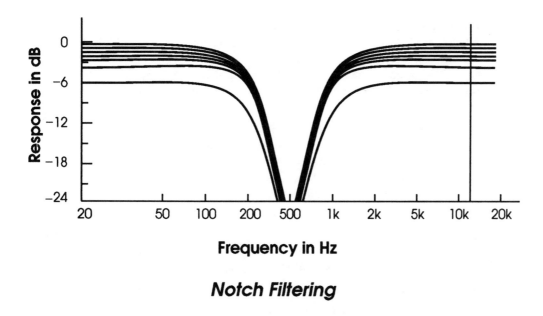

Notch Filtering

Reverberation

Reverberation, or **reverb** as it is commonly known, is often defined as multiple random reflections of a sound wave after the sound source has ceased vibrating. A natural occurrence anywhere sound waves encounter reflective surfaces, reverb increases as the surfaces approach total reflectance (hardness and smoothness) and as the number of nearly parallel surfaces increases. A classic example of reverb is found in a tiled bathroom or shower. This may explain why some people find the shower to be the ideal place to do their singing (a bit of reverb tends to mask a deficiency of natural ability). A room that promotes natural reverb is known as a **wet room,** and one that has soft, absorptive surfaces (drapes, carpet, furniture) that do not reflect sound waves nearly as well is called a **dry room.** Some recording studios can be quickly converted from wet to dry by using changeable panels and movable drapes on the walls. Other audio production areas often attempt to be as dry as possible using specialized acoustic treatments for walls, ceilings, and floors. One such acoustic treatment is Sonex® foam. When an audio producer recording in a very dry room desires a bit of reverb, it is often furnished by the addition of artificial or electronic reverb using an analog or digital reverb unit.

Compression and Limiting

In many broadcast applications, audio is processed through a **compressor** or **peak limiter.** The main reason for this processing is to prevent overmodulation while making sure that the signal is as "loud" as is legally permissible. Both processes tend to decrease the dynamic range of the audio, that is, they reduce the overall amplitude range of the signal.

COMPRESSION: Reduction of dynamic range. Used in broadcasting to achieve greater or more uniform loudness. Digital compression involves the use of algorithms to reduce the bandwidth necessary to store or transmit a digital signal.

LIMITER: An electrical device that prevents an audio signal from superseding a predetermined limit. Used to prevent overmodulation.

The peak **limiter** is actually an amplifier that automatically reduces its gain as the strength of the incoming signal approaches a specified threshold. The idea is to reduce the audio peaks before the signal distorts from overloading. Some audio tape recorders and most videotape recorders have built-in peak limiters. However, they may be fairly unsophisticated and may not offer the control or smoothness required for professional work.

The compressor has a slightly different function. By amplifying the "quieter" parts of the audio signal the compressor creates the perception of increased loudness, but at the expense of reduced dynamic range. Another thing to keep in mind is that a compressor will amplify low-level signals without consideration of whether the signal is desirable. This pumping sound, as the background noise rises and falls, is best avoided by adjusting the settings on the compressor.

Another option is to use the compressor in conjunction with an expander. The expander performs a function that is the opposite of that performed by the compressor. Instead of acting to even out the dynamics of the audio signal, the expander reduces its gain for quiet passages and increases its gain as the signal increases in strength. A less sophisticated expander is the **noise gate**, which essentially cuts off the audio once it falls below a certain level. Noise gates can be helpful when background sounds cannot be eliminated any other way.

COMPANDING: A noise reduction technique that uses a two-stage process: compressing the signal before sending (reducing its dynamic range) and then expanding the signal once it has been received to restore it to its original dynamic range.

Compression is also used as part of a noise reduction process known as **companding**. The signal is first compressed, then transmitted, and then expanded to its original dynamic range. The net result is that the audio signal sounds the same as before being processed but has less noise due to transmission loss.

EXPANSION: The increase of the dynamic range of an audio signal. The opposite of compression.

Another similar sounding but otherwise unrelated process is known as **time compression** or **expansion**. Digital technology has made it possible for an audio signal to be sped up or slowed down without affecting the pitch. This can be especially convenient when attempting to make an audio segment time out to a precise length, e.g., for commercial spot production. However, another popular use for time compression, especially among station owners, is the ability to play back a 30- or 60-minute radio or television program in 28 or 56 minutes or less, allowing the station to sell additional commercial spot time. This is sometimes referred to as *Lexiconing*, so named for Lexicon, the manufacturer of a popular audio time compressor. Of course, for television programs, you must play back the program to be shortened on a VTR that has dynamic tracking—this allows for variable speed playback of the video signal. In this case, time compression is used to reduce the pitch of the accelerated audio.

Noise Reduction

DOLBY: British inventor, whose most notable invention was the Dolby noise reduction system. Dolby A, B, and C, and Dolby SR are all types of electronic processing used to increase signal-to-noise ratio and reduce unwanted frequencies, specifically the tape hiss inherent in the recording process.

The father of noise reduction is Sir Thomas **Dolby**, inventor of the various **noise reduction** systems bearing his name. Dolby® A, Dolby B, Dolby C, and Dolby SR are world famous when it comes to audio signal processing and noise reduction. Another popular noise reduction system is dbx®. The two systems are not compatible and are frequently used for different applications. Many audio professionals agree that dbx is more popular in the music recording business, and Dolby is preferred for TV and film soundtrack work.

Dolby A, used in professional sound recording applications, divides the frequency spectrum into four bands and applies companding techniques to each band separately. Dolby A provides from 10 to 15 dB of noise reduction. Type A noise reduction is commonly used in theatrical film projection. Dolby B is used in FM broadcasting and in consumer audio gear. It provides 10 dB of noise reduction across a single band. Dolby C, commonly used in high-speed audio duplication, reduces noise by 20 dB, again using a single band. Finally, Dolby SR, which stands for spectral recording, is a relatively new development from Dolby Labs. It uses a combination of Dolby A and C technology to achieve excellent noise reduction and greatly increased headroom.

The dbx noise reduction system is available in either the Type I or the Type II configuration. The Type I system is for high-speed tape recording, and the Type II system is designed for slower tape speeds. Like the Dolby B and C systems, both dbx systems reduce noise by companding a single band across the entire dynamic range of the audio signal.

Audio Connectors

An area that usually creates a great deal of confusion for many audio and video professionals is audio connectors. Professional and consumer gear is sometimes used side by side, and the input and output connectors come in every possible configuration. The common audio connectors include **XLR** or **Canon, phone, phono** or **RCA**, 3.5mm **mini**, and the Deutsche Industrie Normen (**DIN**) connectors. There are numerous others. The two major categories of audio connectors are **balanced** and unbalanced. An example of the former is the XLR connector, and an example of the latter is the phono connector. The difference between balanced and unbalanced connectors is readily apparent by noting the number of contact surfaces or pins. A balanced line or connector carries two signals plus the shield or ground, and an unbalanced line has only one conductor plus the shield. It is the addition of the third conductor that makes a balanced line less susceptible to interference and allows longer cable runs. In most instances, balanced lines are used with low-impedance equipment, and unbalanced lines are used with high-impedance gear. However, there does seem to be a trend toward using balanced high-impedance wiring in some applications. One advantage of such a system is the lower crosstalk associated with the Hi-Z signal.

Most professional and broadcast equipment that uses balanced lines has XLR connections. If you need to convert an audio signal from balanced to unbalanced, you should use an audio transformer (a handy item for the remote location audio bag).

Once you've mastered the various types of connectors, you must learn to distinguish between male and female connectors, often referred to as *plugs* and *jacks*. When using XLR connectors, it is generally accepted that male connectors are used for outputs and female for inputs. With phone and RCA connectors, chassis mounts are typically female, and cables are male.

As mentioned above, the standard for professional audio connectors is the XLR, or Canon, connector. The XLR has three pins—two conductors and one ground—and a fourth contact for chassis ground. It is necessary to ensure that they are wired correctly, or they will be out of phase. Actually, what is most important is that you remain consis-

DIN: Deutsche Industrie Normen. The German industry standard for electrical and electronic devices and connectors. European equipment manufacturers often use DIN connectors.

BALANCED: A three-conductor system for carrying audio signals that reduces hum and signal interference over long cable runs. Balanced connectors typically have three contacts, e.g., XLR or Canon connectors.

tent with all of your equipment. The standard in most applications is pin 1-shield (ground), pin 2-high, and pin 3-low. Some studios and equipment manufacturers reverse pins 2 and 3, and that can cause real problems. It's not a bad idea to have handy a few well-labeled XLR adaptors that reverse the second and third pin. These can be used to interface with an audio source or feed that operates on the other standard.

One audio problem common to remote production, especially when using AC power rather than battery power, is improper grounding. This frequently shows up as a 60- or 120-Hz hum or buzz. This can be caused by no ground at all or by a **ground loop**. A ground loop is caused when you have more than one connection to ground. The easiest way to make sure that you have only one ground is to insert a **ground lifter** at one end of the connecting cable. An XLR barrel connector with a disconnected ground (pin 1) makes a convenient ground lifter and should also be a part of your audio bag. Be aware, however, that faulty AC wiring in the wall outlet could be causing the problems, and the solution may be as simple as changing the power source to a different circuit. Circuit testers are available that allow for quick evaluation of an AC outlet's condition. Also, be sure to check the other equipment that is using the same AC circuit. A common wall-mounted light dimmer is one of the worst culprits for causing hum, and some vending machines are also likely suspects.

Note that some consumer video gear uses phono or RCA connectors for both audio and video connections. However, bear in mind that the frequency of the video signal is much higher than that of the audio signal. You can use cable rated for video frequencies, fitted with the proper connectors, in the place of the audio cable, but you should never use a cable rated for audio signals to carry video signals. The high frequencies may not be passed by the audio-grade cable, and the result will be a lower resolution signal.

Patching

In most audio facilities, equipment is not hard-wired, that is, connected directly to other pieces of equipment. Rather, signals are routed through a **patch bay** or **patch panel**. Whether the audio facility is patched physically or electronically, all of the outputs and inputs of each piece of equipment are brought to one central location but not connected. This way, things can be connected, or patched, easily. In the case of the physical patch bay, a short length of audio cabling known as a **patch cord** is used. With electronic patching, or **routing**, equipment, the crosspoints are connected electronically by selecting push buttons at the operator's station. In most cases, patch bays are arranged with the jacks in two rows, outputs on the top and inputs on the bottom. Each jack is carefully labeled as to its source or destination. It is common procedure to arrange each output above the input to which it would normally be connected. The **normaled** patch bay is wired so that the output is directed into the input directly below *unless* the circuit is broken by a patch cord. This arrangement requires the use of patch cords only in situations that require unusual arrangements of equipment and signal flow.

Output

Input

Output ATR

Input Channel 2

Normaled: Terminals
Wired Together

Patch Panel

Self-Study

Questions

1. The amplitude of a sound wave is perceived as loudness, which is measured as:
   a. decibels
   b. amps
   c. pitch

2. The standard audio connector for professional, balanced audio lines is the:
   a. RCA, or phono
   b. phone
   c. XLR, or Canon

3. This type of microphone element requires a supply of DC voltage to perform its transducing function:
   a. condenser
   b. dynamic
   c. ribbon

4. This microphone pickup pattern defines a microphone that is equally sensitive to sounds coming from all directions:
   a. cardioid
   b. omnidirectional
   c. unidirectional

5. The standard line level for most broadcast audio equipment is:
   a. –4 dB
   b. 0 dB
   c. +4 dB

6. This digital audio format uses a 4.7-inch optical disk:
   a. compact disk
   b. DAT
   c. laser disk

7. This type of audio equalizer allows the operator to adjust the Q factor, which is the width of the bandwidth to be affected:
   a. parametric equalizer
   b. graphic equalizer
   c. selectively variable equalizer

8. Variations in pitch due to instability in tape or disk speed is known as:
   a. S/N ratio
   b. pink noise
   c. wow

9. Ribbon microphones naturally have this type of polar pattern:
   a. omnidirectional
   b. bidirectional
   c. cardioid

10. Dolby and dbx are trademarks for systems that can be used to increase the S/N ratio of audio recordings. Increasing the S/N during the recording process is more commonly known as:
    a. graphic equalization
    b. companding
    c. noise reduction

Answers

1. a. Yes. Decibels are the standard unit of measure for the loudness of an audio signal.
   b. No. Amps are a measure of electrical power, not loudness.
   c. No. Pitch has to do with frequency or tone.

2. a. No. The RCA, or phono, connector is favored for consumer equipment applications.
   b. No. However, phone connectors are sometimes used in professional audio applications.
   c. Yes. The three-conductor XLR, or Canon, connector is preferred for most broadcast applications.

3. a. Yes. The condenser microphone element requires a small amount of DC voltage, usually supplied by either an onboard battery or phantom power.
   b. No. Dynamic microphones do not require an external power supply.
   c. No. Like the dynamic microphone, the ribbon mike does not require external DC voltage.

4. a. No. The cardioid pickup pattern is heart shaped, which defines a microphone that is more sensitive to sounds coming from the front.
   b. Yes. The prefix *omni* means *every* or *all,* as in *all directions.*
   c. No. In this case, the prefix should make the answer obvious. *Uni* means *one,* as in *sensitive to sound coming from one direction.*

5. a. No.
   b. No. This is lower than standard line levels for professional equipment.
   c. Yes. This is typically the standard line level for professional audio gear.

6. a. Yes. The CD is becoming widely used as a playback-only medium for broadcast use and is beginning to be used for recording as well.
   b. No. Digital audio tape is a magnetic tape medium.
   c. No. Laser disks are an optical medium, but they are used for audio and video and are larger in diameter than CDs.

7. a. Yes. The parametric equalizer is the only equalizer that allows operator control of the Q factor.
b. No. Graphic equalizers usually come preset with a fixed Q factor.
c. No. Selectively variable equalizers allow the operator to select the frequencies to be affected, but not the width of the band to be affected.

8. a. No. S/N has to do with the ratio between the signal level and the noise level.
b. No. Pink noise is a test signal used to analyze audio components and acoustics.
c. Yes. The word *wow* is onomatopoeic—it sounds like it sounds.

9. a. No. Ribbon microphones are seldom omnidirectional.
b. Yes. Ribbon microphones are by design bidirectional, although they can be made to have other pickup patterns.
c. No.

10. a. No. The graphic equalizer is used to boost or cut selected frequencies, not to alter the signal-to-noise ratio.
b. No. Companding is part of the noise reduction process, but it is not the name of the process.
c. Yes. Increasing the signal-to-noise ratio by employing Dolby or dbx technology during the recording process is known as noise reduction.

Projects

Project 1

Collect as many audio connectors, cables, and adaptors as possible. Look at the way that the various connectors are wired.

Purpose

To learn the various types of audio connectors used in a broadcast production facility and to be able to distinguish the differences between balanced and unbalanced connectors.

Advice, Cautions, and Background

1. Try to locate both male and female connectors for each type.

2. Molded connectors cannot be disassembled without destroying the connector and cable. You may need a small screwdriver to disassemble the XLR connectors. Make sure that you do not loose any pieces, and reassemble the connectors when finished.

How to Do the Project

1. After you have gathered as many types of connectors as possible, categorize them according to the following criteria:
- consumer vs. broadcast
- male vs. female
- balanced vs. unbalanced
- two-wire vs. three-wire cables and connectors

2. List the various pieces of audio/video equipment in your facility and the type of audio connectors required by each.

3. Submit both lists to your instructor to receive credit for this project.

Project 2

Visit a local audio production facility or radio station.

Purpose

To see and hear the various types of outboard audio equipment usually found in professional audio facilities.

Advice, Cautions, and Background

1. Try to find a place that has the following audio equipment:
- equalizers (more than one type, if possible)
- compressors and expanders
- reverb and delay
- harmonizers, time compressors, and expanders

2. Remember as you listen to the audio effects that the "wildest" sounding effects probably have the most limited application. More often than not, the audio engineer uses just enough of one or more effects to achieve a unique sound or to fix a problem but not enough for the listener to be aware that the audio signal is being manipulated.

3. Be punctual, be prepared with appropriate questions, and be careful that you do not interfere with normal working operations.

How to Do the Project

After the trip, prepare a short paper describing the various audio devices you observed and a short explanation of how those devices might be used in the audio production process.

Project 3

Categorize various microphones.

Purpose

To become familiar with the microphones in your facility—their type, pickup pattern, and application.

Advice, Cautions, and Background

1. Be sure to handle microphones gently; some can be quite fragile. Return the microphones to the place where you found them when you're finished.

2. Most of the information about a microphone can be gathered by looking carefully at its case or housing. Is there a polar pattern inscribed somewhere? Is the microphone designed to be mounted on a stand or fish pole or clipped to an article of clothing? Does it have a place to insert a battery? Take note of the proper position of the battery with regard to the positive and negative terminals.

How to Do the Project

Make a chart listing the various microphones by their manufacturer and model number, followed by the various attributes. Submit the list to your instructor to receive credit for the project.

Chapter 3
VIDEO

NTSC: National Television Systems Committee. The committee formed to determine the guidelines and technical standards for monochrome and, later, color television. Also used to describe the 525-line, 59.94-Hz color television signal used in North America and several other parts of the world.

LUMINANCE: The brightness information part of the video signal. Luminance is often designated by the symbol Y. On a waveform monitor, the luminance level of the video signal can easily be measured by viewing in the L-Pass or IRE mode.

CHROMINANCE: The color information part of the video signal, usually defined in terms of hue and saturation. The video signal is made up of chrominance (color) and luminance (brightness) information.

The standard for color television in the United States is known as **NTSC**, the initials of the **National Television Systems Committee**. Years of planning preceded FCC approval of this system, which allowed complete compatibility with monochrome receivers while delivering brilliant color to those who purchased color receivers. However, to put things in perspective, remember that television engineers lovingly refer to NTSC as "Never Twice the Same Color." One reason for this is that NTSC is the only television standard that allows the viewer to adjust the hue of the television receiver's picture.

In basic terms, the NTSC composite video signal can be understood as a **luminance** signal and a **chrominance** signal combined to provide both brightness and color information. Luminance information alone is all that is necessary for a black-and-white or monochrome image. The addition of the chrominance signal, which contains information about hue and saturation, fills in the necessary color information. Movement is achieved by a scanning process that presents a new frame of information every 1/30 of a second. Before going too far, it would be helpful to look at the source of video signals. Although not the only one, by far the most common source of video signals is the video camera. A great variety of cameras are available today, offering a wide range of price tags, features, and image quality. The place to begin is with the video camera and the origin of the video signal.

The Video Camera

The purpose of the video camera is to convert an optical image into an electrical signal that can then be converted by a video monitor into a visible (or optical) image. The video camera is made up of several essential parts and various optional parts that are determined by its intended use or application. Two of the essential parts are the optical element or elements and the imaging device or devices. The primary optical element is the lens, and its function is to gather and focus the light reflected from the subject onto the face of the imaging device. A lens may be made up of few or many glass elements, it may have a fixed or a variable focal length, and it may offer variable or fixed focus. In any case, it should be understood that the image quality of the resulting video begins with and is greatly affected by the lens.

PICKUP TUBE: The former standard (see also CCD) for imaging devices in video cameras. A pickup tube is a transducer that converts light in electronic energy by means of an electron beam that scans a photosensitive faceplate. The most common tubes are the Saticon® and Plumbicon® tubes.

CCD: Charge-coupled device. These solid-state imaging transducers are used in video cameras in place of pickup tubes. They provide numerous advantages over tubes and are expected to replace tubes entirely within a few years.

The imaging devices are transducers that convert light into electrical signals. The two primary types of transducers are **pickup tubes** and **charge-coupled devices (CCDs)**. Tubes and **chips**, as CCDs are commonly called, will be discussed in more detail later in this chapter. A camera may have only one transducer or as many as four. For broadcast video cameras, the norm is three, one for each of the three primary colors for video: **red**, **green**, and **blue**. Three-tube or three-chip cameras, then, must have something between the lens and the imaging devices to split the light into its red, green, and blue components. The most common beam splitter is the prism, an angular block of optical-quality glass that directs the various lengths of light waves into different directions. Once the light has been directed onto its appropriate tube or chip, the signals from each tube or chip can be combined to reproduce the full spectrum of color. The luminance information, incidentally, is composed primarily of the signal from the green channel.

The Camera Chain

In a studio configuration, a video camera has several components that together make up the camera chain. These include the **camera head**, the **camera control unit (CCU)**, the **remote control unit (RCU)**, the **sync generator**, and the **power supply**.

The imaging devices are housed within the camera head, and that is where the electronic image is created. Typically, the camera head has several important attachments, including the lens, a viewfinder, and operator controls for zooming, panning, and tilting.

The CCU contains the camera's encoding circuitry. The function of the RCU is to permit remote control of the camera's CCU by a video engineer. The video operator sits in front of the RCU and "tweaks" the camera, adjusting the iris, video and black levels, and color balance. The camera must be set-up before the shoot and must be continually adjusted during the shoot to ensure that multiple cameras match, or look alike in terms of exposure and color reproduction. This process is commonly known as *shading* or *painting*. The RCU allows these adjustments to be made in the control room instead of at the camera head. It is important to note that during all of these adjustments, the video engineer relies on the waveform monitor and vectorscope in addition to the picture monitor to make the adjustments.

The **sync** generator drives the camera head so that its signal is in synchronization with other video signals in the control room. The synchronization pulses perform a function much like that of an orchestral conductor whose responsibility it is to keep the orchestra playing in time. Stable sync is the heartbeat of the technical facility, and it is what permits many pieces of video equipment to operate together without glitches, rolls, or shifts. In ENG cameras, the sync generator is built into the camera, and the portable video cassette recorder (VCR) operates in sync with the camera. In a studio with many cameras, switchers, and VTRs, one central source of sync drives the whole system. This is typically called a **reference sync generator** or **master sync generator**. Sync generators in other pieces of equipment are locked, or **genlocked**, to master sync so that all video signals are in sync with each other.

The power supply simply provides AC power to the camera head and CCU. Studio camera heads are supplied with AC power through multiconductor camera cable that connects the camera head and CCU. ENG cameras can be supplied with either AC or battery power through the VCR and camera cable or they can be powered by their own external battery pack.

Monochrome and Color

Monochrome, which literally means *one color*, is the designation for a television display that reproduces the brightness value of the scene without color information. A monochrome camera needs only one tube to produce a black-and-white (and shades of grey) picture. For certain applications today, in science and surveillance for example, a monochrome image is not only acceptable, but may actually be desirable for its higher resolution image.

CCU: Camera control unit. This is the part of the camera chain that houses many of the control and setup functions of the camera. Usually located apart from the camera head (in the control room or remote truck), the CCU allows the cameras to be shaded during a real time multicamera production.

SYNC: Short for *synchronization pulses*. Horizontal and vertical sync pulses drive the scanning process at the camera and picture monitors. Composite sync (H and V sync combined) is sometimes required by video components to interface properly with other components. A sync generator is a master source of sync for a facility.

GENLOCK: The locking of a video component or system to the synchronization pulses of an incoming signal. For example, a camera can be genlocked to a switcher or to another camera so that the signals from each can be mixed without any disruption.

PIXEL: Picture element. The smallest unit of measurement when dividing an electronic image. Commonly used to define resolution of imaging or display devices.

As mentioned earlier, to produce a color picture, the light gathered by the lens must first be split into its three primary colors—red, green, and blue. Each of these colors is converted into an electronic signal relating to the brightness value for each **pixel** (short for *picture element*) on the image area. Later, when these three signals are combined, the full spectrum of color and brightness is reproduced.

Some inexpensive and consumer color video cameras use a single tube or CCD to produce a color video signal. In this case, the single tube or chip does the job of three but not without sacrificing image quality. The pickup device is striped with the three primary colors, and each stripe sends a separate signal to the red, green, or blue channels. Because the face of the tube is divided among these three colors, the resolution of the resulting image is considerably lower than with a three-tube or three-chip camera. Its advantages are lower cost, smaller size, and no need for alignment. As a compromise, some manufacturers have released "prosumer" grade camcorders, which use two CCDs—one chip for luminance information and the other for chrominance.

Beam Splitters

As mentioned earlier, a three-tube or three-chip camera requires some sort of beam-splitting device between the lens and pickup element. Although two options were once available (**dichroic mirrors** and **prism block assembly**), only the latter is considered to be a viable option today. Dichroic mirrors have been superseded by the prism block assembly and are no longer in production. One of the problems with the dichroic mirror approach was that the mirrors got dirty or were jarred out of alignment; both cases required more frequent servicing for cameras with dichroic mirrors than those with prism blocks. In the case of CCD cameras, another advantage of the prism block is that the chip

PRISM BLOCK: This piece of optical glass acts as a beam splitter in a three-tube color video camera, separating the light gathered by the lens into its red, green, and blue components. The prism block has virtually replaced the dichroic mirror method of beam splitting.

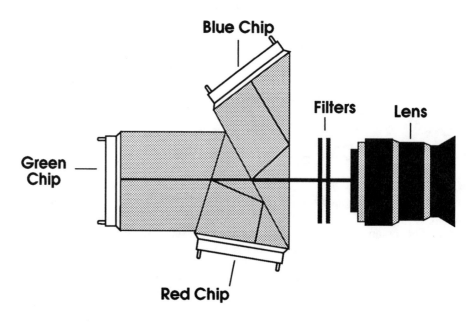

Prism Beam Splitter

is cemented directly to the prism. This creates one solid unit that requires none of the ordinary maintenance required to position the pickup devices in relation to the red, green, and blue beams of light.

Imaging Devices

There are currently two categories of imaging devices: **tubes** and solid-state CCDs. Of these, tubes are being phased out in favor of CCDs at a remarkable rate. Although tubes continue to be the imaging device of choice for a few specific applications, CCDs have quickly moved from infiltrating the inexpensive ENG camera market to invading the high-end studio camera market, and there appears to be no stopping them until every tube camera has eventually been replaced. Before the advantages of CCDs are discussed, the various types of tubes and their characteristics are explored.

Tubes

Camera tubes come in a variety of types and sizes, but by far the most common types are the **Saticon**® and **Plumbicon**® tubes in 2/3-inch size. Saticon and Plumbicon are trademarks of Hitachi and Philips, respectively, and the names provide clues about the composition of the photosensitive material used in the construction of the tubes. Saticon tubes contain the chemicals selenium, arsenic and tellurium, and Plumbicon tubes are manufactured out of a lead-based material (Pb is the symbol for lead). Other variations of video imaging tubes include Newvicon® (Matsushita) and Trinicon® (Sony).

Camera tubes range in size from 0.5 to 1.25 inches in diameter (13 to 30 mm). Generally speaking, the larger the tube, the higher the **resolution**. Resolution, or picture sharpness, might best be visualized by comparing the sharpness of a glossy magazine photo to a newspaper photo. The magazine photo has much higher resolution, resulting in increased detail and sharpness.

Some attributes of the ideal camera tube include excellent resolution; high sensitivity in low light; resistance to burn-in, lag, and comet-tailing; accurate reproduction of color; and low cost. Different types of tubes achieve these goals with varying degrees of success. The Vidicon tube was one of the earliest models on the scene and is considered to be the progenitor of those that have followed. The newer tubes are simply improvements of the same basic technology used in the design of the Vidicon tube. In the industrial and professional markets, the Saticon and Plumbicon tubes are the two leaders, with broadcasters showing a slight preference for the latter. However, the advantages of Plumbicon tubes are realized at a price; they are considerably more expensive than other types of tubes.

Some of the differences between Plumbicons (plums) and Saticons (sats) include these: plums generally cost more; plums are more sensitive to light (1 to 1.25 f/stops); plums are chemically treated to react to either red, green, or blue light and should therefore be replaced by tubes with the same classification; Plumbicons tend to be more easily damaged by burn-in (making them a more practical choice for studio cameras than for industrial or ENG applications); and finally, plums generally have slightly lower resolution than saticon tubes do. Both Saticon and Plumbicon tubes are available in **diode gun** versions. These tubes have increased resolution over standard tubes. Diode gun Saticons can resolve over 2000 horizontal lines, far more than any tape machine can record or, for that matter, nearly six times the resolution capability of a broadcast television signal.

RESOLUTION: Generally defined as the ability to convey detail or the number of pairs of black/white lines that can be distinguished. Camera imaging devices and CRTs exhibit varying degrees of image sharpness. Likewise, video recording devices may record and play back with similarly varying results, which are largely dependent on the frequency of the recorded signal. Engineers typically use a ballpark figure of 80 lines of horizontal resolution for each MHz of bandwidth. Using this equation, the 4.2-MHz bandwidth of a broadcast video signal allows approximately 330 lines of resolution. A camera's resolution is usually defined by its horizontal resolution, i.e., the number of vertical black/white line pairs that it is capable of resolving. An EIA RETMA resolution chart is commonly used to measure a camera's resolution capability.

Some common problems associated with tubes include image retention (similar to human persistence of vision), fragility, and a need for periodic registration adjustments.

Caring for a burned tube

BURN (BURN-IN): A flaw in a camera pickup tube that results from extended overexposure. CCDs, unlike tubes, are immune from burn-in.

Burn-in is caused when a tube camera remains focused on a very bright or high contrast scene for a length of time sufficient to make a lasting impression or image on the face of the pickup tube. A minor burn-in will heal itself in a short time (ranging from a few seconds to a few hours). However, a more severe burn-in may require corrective steps. One way to reduce the effect of a burn-in is to expose the tubes to an evenly illuminated white card. It is recommended that the camera lens be adjusted out of focus to make the image softer. Leaving the camera tube exposed to this bright, even illumination for a few hours or overnight may heal or reduce all but the most severe burn-ins. However, be aware that this process will likely shorten the life span of the tubes.

Charge-Coupled Devices

Charge-coupled devices (CCDs), unlike tubes, are solid-state imaging devices. They are small, lightweight, and totally resistant to lag and burn-in. They require less power to operate than tubes, they are very rugged, and theoretically at least, they should last for the life of the camera—unlike their tube ancestors. The only disadvantages of chips are their lower resolution compared to the best tubes, vertical smear, and something called *fixed pattern noise*. The resolution problem becomes less of an issue, however, as chip technology improves. Another important factor worth considering is that a tube camera that is even the slightest bit out of registration quickly drops below the resolution of a good chip camera. In other words, unless you have an impeccably maintained tube camera in perfect operating condition and are shooting in controlled lighting, today's chip cameras are likely to equal or beat the tube camera's performance. Vertical smear, which appears as vertical streaks of light radiating from highlights, has been greatly reduced by the introduction of frame interline transfer (FIT) chips. Fixed pattern noise gives an appearance as though the camera were shooting through a dirty lens or through a wire screen. The effect is most noticeable when the camera moves, e.g., pans left or right, on a static subject. As the resolution of CCDs continues to improve, this problem is diminishing.

The three types of CCDs in use today are the interline transfer (IT), frame transfer (FT), and frame interline transfer (FIT) chips. The IT chip transfers the image being photographed on a line-by-line basis during the horizontal blanking interval. One unfortunate by-product of the IT chip is potential smearing of the image. The FT chip transfers the image at a frame rate, thereby preventing some of the smear problem. However, the FT chip tends to exhibit poor sensitivity to the color blue. The third type of CCD, the FIT chip, is presently the state-of-the-art in CCDs. It combines the best of both the IT and the FT chips and is found in the higher priced cameras, both field and studio models.

Unlike tube cameras, CCD cameras do not use electron beam scanning. Each CCD is actually a field of discrete semiconductors; each separate transistor corresponds to a pixel. The number of pixels determines the resolution or sharpness of the image reproduced by that CCD. Once the transistors convert the light into an electrical charge, the signal is momentarily stored and read out line by line, or at a frame rate, to conform with NTSC's 525/30 scanning rate.

58

Granted, it is difficult to make a totally flawless CCD when some have over 400,000 transistors on an area 1/2 inch square. After the manufacturing process, quality control inspectors check for defective transistors. Each defect will appear as a tiny white spot on the picture. After a certain number of defects are detected, a chip is rated to be sold as an industrial-grade chip rather than a broadcast-grade chip. Chips with an even greater number of defects are sold to be installed in consumer cameras.

Operating Light Levels

Not too many years ago, studio cameras required between 250 and 300 foot-candles (fc) to make good pictures. Such high light levels were uncomfortable for the talent and required great amounts of electrical power, which in turn made additional air conditioning essential. However, that was the price to pay for decent video in those days. Taking a video camera on location meant shooting in daylight or setting up a considerable number of portable lighting instruments. Much has changed since then. Today's studio cameras frequently operate at 75 to 125 fc and make great pictures at those levels. Portable cameras used for ENG make an acceptable picture with much less light, some going as low as 5 fc. Using the gain switch allows ENG photographers to shoot at night in available light, although the results are seldom impressive. Sony's new HyperHAD CCD (HAD stands for hole accumulated diode) is said to provide a full stop of increased sensitivity. The HyperHAD CCD has a coating of microscopic lenses that concentrate the light onto each of the 400,000 pixels on the face of the chip. The increased sensitivity is comparable to increasing film sensitivity from a speed of 160 to 320 ASA.

Contrast Ratios

Under optimum conditions, a professional-quality video camera, using either pickup tubes or CCDs, can handle, or reproduce, a contrast ratio of nearly 30:1, that is, the brightest spot in the scene can be 30 times brighter than the darkest spot and you will still be able to see detail in both the light and dark areas. Any area of the scene being shot that is either brighter or darker than the 30:1 ratio allows will show up as either completely white or black and will be void of texture. Film, for the sake of comparison, is capable of approximately 100:1, that is, the brightest area of the frame can be 100 times brighter than the darkest area while retaining texture or detail in both areas. This advantage provided by film makes it the choice when extremely high contrast or uncontrolled lighting conditions exist. However, if lighting can be controlled and you are able to bring the contrast ratio within the 30:1 range of video, the resulting image contrast will be comparable.

A classic example of high contrast that can cause problems for video cameras is the male award recipient's black tuxedo and white shirt. If you adjust the iris for proper exposure of the shirt, the jacket goes completely black without any texture, and the skin tones appear dark. If you expose for the jacket, the shirt blooms, and the skin tones are too light. The video engineer must adjust the camera to achieve the most pleasing balance. Some of the newer cameras have built-in compression circuits that can handle a wider range of contrast. However, whenever possible, this type of excessive contrast should be avoided by controlling wardrobe, set, and lighting. Using a "TV grey" shirt would be one way to minimize the problem in this particular situation. The grey shirt will actually appear white on camera once the camera is exposed for proper skin tones.

Preamps and Proc Amps

Once the light entering the camera's lens has been split into its red, green, and blue components, which in turn are converted into electronic signals by the camera tubes or chips, the electronic process continues in the production of the NTSC video signal. The signal generated by the tubes or chips is extremely low level, so low in fact that it must first be amplified by a preamplifier. The preamp must be located very near the pickup device so that electronic noise and electromagnetic interference do not degrade the video signal. This special low-noise preamp must be carefully protected from outside interference signals, e.g., RF, and electrical radiation. The next amplifier that the video signal encounters is the processing amplifier (proc amp). Each channel of a three-tube color camera has a proc amp, which allows adjustments to be made to that particular signal. These adjustments control white balance, black balance, gamma manipulation, and video gain. The proc amps may be contained in the camera head, or they may be in the CCU.

Gain Switch

SENSITIVITY: The measurement of a video camera's ability to produce an acceptable picture in low-light conditions. Increased sensitivity is usually achieved by improvements to the lens quality, imaging device construction, and preamplifier circuitry. The sensitivity of a video camera can be artificially increased by engaging the gain switch. In effect this increases the output of the preamps, amplifying the video signal and any electronic noise.

Most video cameras have a **gain switch**, which permits the user to select +9 or +18 dB gain. Activation of the gain, or **sensitivity**, switch allows the camera to make pictures in low light, but you must be aware that you will sacrifice quality in the process. The gain switch is actually an amplifier that boosts not only the video signal but also the residual noise in the signal. The result is a corresponding reduction in the S/N ratio. Although the gain switch is a valuable asset when shooting in low-light conditions, it should be used only after other options have been exhausted. The rule of thumb is to use the gain switch only in hard-news situations, when it is the only way to make usable pictures. The gain switch should never be used in place of lighting. Some of the newer CCD cameras actually have a gain reduction option. By setting the switch to −6dB you can further reduce video noise. In addition, decreasing the camera's sensitivity allows greater control over the camera's depth of field for creative selective focus.

NTSC's Odd-Even Interlace Scanning

FRAME: One complete video picture. In NTSC, takes place in one-thirtieth of a second and is made up of 525 lines and two fields.

NTSC VIDEO

1 field=
1/60th second, 262.5 scan lines

1 frame=
1/30th second, 525 scan lines*

Although each video frame has 525 lines, 40 lines are used by the vertical blanking interval. That leaves a potential 485 "active" video scanning lines.

To understand the history behind the scanning process for video, it may be of benefit to consider the technology behind motion pictures. Motion picture technology, which preceded the arrival of television technology by nearly 50 years, is credited with the discovery that it is necessary to project an image about 48 times per second to alleviate flicker. Flicker, visible in very early experimental motion picture footage, is perceived as the absence of fluid motion. It is caused by a **frame** rate slow enough for the eye to perceive individual frames. Anyone who has seen early motion picture footage knows that flicker makes viewing for any length of time tiresome. It was not long before early filmmakers found that it was much simpler and more economical to project at 24 frames per second (fps) and use a double-bladed shutter to display each frame twice, thus increasing the apparent temporal resolution and achieving an effect similar to projection at 48 fps. When TV was invented, it was designed to operate at 60 cycles (standard AC power frequency) or 60 frames per second. Unfortunately, it was not practical to broadcast 60 complete frames each second due to limitations of the allocated bandwidth. So,

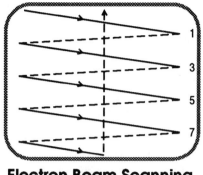

**Electron Beam Scanning
Odd Lines
(First Field)**

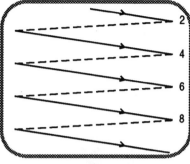

**Electron Beam Scanning
Even Lines
(Second Field)**

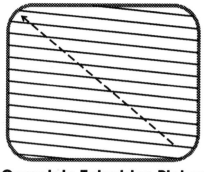

**Complete Television Picture
(Frame)**

Interlace Scanning

television pioneers borrowed an idea from film. Instead of projecting each frame twice, like film projection does, the video engineers decided to divide each frame into two fields, each **field** having half the resolution of the complete frame. Instead of 60 discrete frames making up one second of video, there are 30 discrete frames, and each frame is divided into two fields. Because the video frame is made up of 525 horizontal lines, displaying 262.5 lines twice each thirtieth of a second results in only slightly diminished picture resolution with greatly increased temporal resolution. The result is a flickerless 60 fields per seconds and fairly smooth on-screen movement.

However, unlike film, the video image is not created, stored, or displayed an entire frame at a time. Rather, each frame of video is broken down into a continuous stream of information. This is achieved by a scanning process that begins in the camera's imaging device and is completed on the screen of the video monitor's picture tube. At the camera head, the optical image is scanned line by line, from the top of the frame to the bottom, over and over again. First, all the odd-numbered lines are scanned (making up field 1), and then all the even-numbered lines are scanned (field 2). This process is repeated for each new frame. The actual **odd-even interlace scanning** process uses an electron gun, which directs an electron **beam** at a target, also known as the **raster**. The beam begins by scanning line number 1 from left to right. Once it reaches the end of line number 1, it shuts off and drops down to scan line 3. This process continues until it has scanned all of the odd-numbered lines. At this point, the beam shuts off and retraces back to the top of the raster to begin scanning line number 2. After the even-numbered lines have been scanned, the process begins again with the odd-numbered lines of the second frame. Remember, this process of first scanning the odd-numbered lines (field 1) and then the even-numbered lines (field 2) takes place 30 times every second.

A very important part of the scanning process has to do with the synchronizing pulses that tell the electron gun where to aim, when to shoot, and when to turn off. These signals are known as **horizontal** and **vertical sync pulses** and **blanking**.

FIELD: Half of a video frame, 262.5 horizontal lines (NTSC).

ODD-EVEN INTERLACE SCANNING: The method by which imaging devices and picture tubes create and display 262.5 lines of information 60 times each second. An electron beam scans all of the odd-numbered lines first (one field) and then scans all of the even-numbered lines (the second field), thereby creating an entire frame (two fields) composed of 525 scanning lines. It repeats this process 30 times every second to produce a flickerless image.

BEAM: The focused stream of electrons that scans the face of the camera's pickup tube and the monitor's picture tube.

RASTER: The portion of the camera pickup tube or CRT that is traced by the scanning electron beam.

BLANKING: The portion of the video signal that turns off or *blanks* the scanning beam during retrace. Blanking has both vertical and horizontal components. On a waveform monitor, horizontal blanking is the signal information between active lines of video.

Horizontal sync (H sync) is a 15,750-Hz signal that tells the electron gun to retrace after the beam scans each of the 525 lines (this takes place at the end of each line), 30 times a second. The frequency of H sync is easily computed by multiplying 525 lines by 30 frames per second ($525 \times 30 = 15{,}750$).

Vertical sync (V sync) is a 60-Hz* signal that tells the electron gun to retrace at the end of each field. This happens twice for each frame (two fields per frame), and you recall that there are 30 frames per second. By multiplying 30 by 2, we get 60.

The combination of horizontal and vertical sync is known as **composite sync,** or just **sync,** and is an important signal for most television equipment, especially when it is used in a studio or editing configuration where signals from various sources are combined and used together. Today, much of the broadcast equipment that requires a composite sync signal can use **color black** or **black burst**. Black burst is composed of composite sync, reference burst, and a black video signal that is normally at 7.5 IRE units above blanking. (Video levels are commonly referenced to IRE [Institute of Radio Engineers] units. The waveform monitor uses a scale spanning more than 140 IRE units. The waveform monitor and the IRE scale are described in detail in the following chapter.)

Horizontal and vertical blanking are the signals that tell the beam to turn off for the duration of the retrace period so that the television monitor will not receive a picture during the time that the electron beam is retracing to begin a new line or a new field. The

BLACK (COLOR BLACK or BLACK BURST): A composite color video signal made up of composite sync, burst, and a black video signal, normally at 7.5 IRE units above blanking.

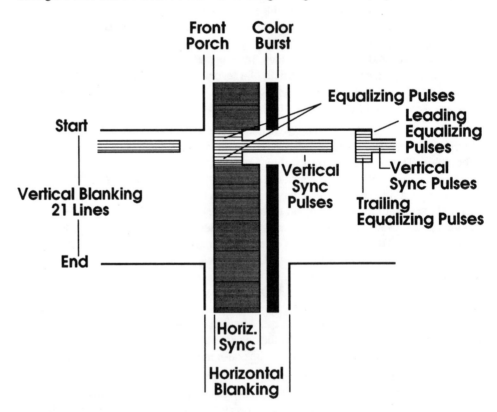

Horizontal and Vertical Blanking Signals

*Actually, the V sync rate is 59.94 Hz, but 60 is close enough for now.

vertical blanking interval can be seen when the TV picture rolls—it's the space between the bottom of one frame and the top of the next. The horizontal blanking interval is the area off to the left or right of the visible frame. The horizontal blanking interval contains the horizontal sync pulse and color burst.

Other Color Television Standards

As mentioned earlier in this chapter, NTSC is the standard television system in use in the United States, in much of Central America, and in parts of the Far East. Other, perhaps technically superior, systems are used throughout the rest of the world. Phase alternation by line (PAL) and sequential color and memory (SECAM) use a different frame rate (25 fps) and have more scanning lines (625) than NTSC. Programs produced in one system must be converted to be broadcast over another transmission system. Recent improvements in digital standards converters have reduced much of the motion artifacts that have always been a problem with standards-converted video.

Encoding

When the standard for the present color television signal was being established by the National Television Systems Committee back in 1953, one decision that was to have lasting repercussions was to make sure that color broadcasts would be compatible with monochrome television receivers. That one decision has resulted in some rather interesting developments in the way that NTSC video is created, recorded, and transmitted. **Composite** NTSC video uses a system that creates a red, green, and blue reproduction of the image. These separate images are then combined into luminance and chrominance signals, which are then combined before the signal is broadcast. The combining of signals is known as **encoding** and is performed by an encoder.

The video signal is made up of two different signals: the chrominance, or color, information (hue and saturation) and the luminance, or brightness, information. Although chrominance information is composed of equal amounts of signal from each of the tubes in a three-tube color camera, the luminance signal is composed primarily from the green tube's signal. The ratio of composition of the luminance signal is 30% red, 59% green, and 11% blue. This is for two reasons: first, the green channel optics are the most direct, and second, humans are most sensitive to brightness values in the green range of the spectrum. Early on, it was discovered that humans do not perceive sharpness nearly as well when looking at color, or chrominance, as they do when viewing brightness, or luminance. Therefore, it was found to be possible to superimpose a low-resolution color signal over a high-resolution monochrome signal. The viewer's perception is of a high-resolution color picture. As mentioned earlier, composite NTSC video encodes the chrominance and luminance information, the result being a low-resolution color signal with the detail provided by the luminance information. This encoding process is what makes color NTSC compatible with monochrome monitors. It's also what makes the video signal compact enough to fit into a single coaxial cable and use less frequency

COMPOSITE: Standard video that combines chrominance and luminance information by encoding the output of the red, green, and blue channels into the Y, I, and Q signals. Composite video also includes blanking and sync and is the standard for broadcast transmissions of video signals.

ENCODE/DECODE: The process of converting video from its RGB components into composite video, and vice versa. Great improvements have been made in the area of encoding and decoding by Faroudja Laboratories, the result being improvement in the quality of encoded NTSC video.

RGB: Red, green, blue. Video in its purest form, as divided into its red, green, and blue components. Some computer graphics devices and monitors work with an RGB signal, but it is not a common format for recording or transmission technology.

bandwidth for transmission. To transmit an unencoded red-green-blue (**RGB**) signal requires three wires and nearly twice the bandwidth necessary to carry an NTSC signal. A closer look at this encoding process is in order.

As previously mentioned, NTSC video does not use an RGB signal to transmit color information. Rather, it converts the three color signals (red, green, and blue) into a luminance signal (Y) and two color-difference signals (R-Y and B-Y). Because the luminance signal is made up primarily of the green signal, all three color signals can be electronically calculated given the luminance and color-difference signals. In addition, in the **I/Q encoder**, the B-Y signal is converted to the **inphase subcarrier (I) signal**, and the R-Y signal is converted to the **quadrature subcarrier (Q) signal**. Both of these signals are reduced in frequency (and therefore sharpness), but remember, the apparent resolution or sharpness comes from the Y signal. Summing the I and Q signals results in the chrominance signal (C). Video recording formats that record and play back the video signal in the Y/C mode include S-VHS® and Hi-8®. Summing the Y and C signals results in NTSC or, in other systems, PAL or SECAM.

Despite its many amazing technical achievements and the simplicity provided by a one-wire system, composite video, as described above, has several shortcomings. These encoding artifacts include moiré effect, cross-color artifacts, smeared colors, and chroma crawl. All of these problems are a result of interference between the encoded chrominance and luminance signals. Almost since the invention of NTSC composite video, there have been attempts to correct some of these inherent problems, which incidentally are often most severe in the recording and playback process.

Recently, some improvements have been realized by the **component** video recording formats. The new 1/2-inch professional video recording systems, Betacam SP and M-II, use this component recording system, which records luminance and chrominance information on separate channels. This technology allows 1/2-inch videotape recording technology to approach and even exceed the quality of 1-inch type C video recording. Betacam and M-II VTRs record the Y, B-Y, and R-Y signals on adjacent tracks on the videotape, thus allowing them to remain separate during the recording and playback process. Full component suites using component switchers, component digital effects units, component character generators, etc., allow the chrominance and luminance information to be kept separate throughout the recording, editing, and processing stages until the last possible moment. Finally, a high-quality encoder converts the signal to composite video before being broadcast or recorded onto a composite video recording format. Some presentation video monitors are now being manufactured that accept a component video signal, avoiding some of the problems that are encountered when displaying composite video.

COMB FILTER: An electronic filter used to lessen the effects of cross-color, cross-luminance interference in the video signal.

Another approach to solving these problems is presented by Faroudja Laboratory's **SuperNTSC**™ system. Using sophisticated **comb filtering** in the encoding and decoding process, the Faroudja circuitry has provided spectacular results from an otherwise standard NTSC signal. SuperNTSC is not considered to be an HDTV system, but the manufacturer hopes it will become an intermediate step for the conversion from NTSC to the accepted HDTV standard of the future.

Video Cable

The cable used to carry a video signal must be capable of handling a very wide bandwidth, high-frequency signal. There are several types of video cables, which are used for different applications. Probably the most common is the **coaxial,** or **coax cable.** This cable is made up of a core, either solid or stranded, surrounded by insulation and a braided shield. The outer layer is plastic or rubber insulation, usually black but recently available in almost any color. Usually, only one video signal is sent over each cable. A familiar grade of coax cable is RG-59 which is commonly used to connect video components in a studio control room or editing suite. The connections required to route video from a switcher to a VCR or from a camera CCU to a monitor or from one VCR to another will commonly be made using coax cable.

Another type of cable used to carry video signals is called **triaxial,** or **triax, cable.** The core is surrounded by two shields. This configuration allows multiple signals (multiplexed) to be sent over a single cable. This is a fairly expensive way to have remote control of a camera using a CCU, but it is much less cumbersome than the traditional multicore camera cable, and it can be run for a much greater distance. A typical application for triax cable is to provide CCU control of fixed (hard) and hand-held (soft) cameras on large multicamera remote productions, such as sporting events.

A fairly specific task is handled by the **multicore camera cable.** This is a large-diameter cable that has many single cables inside. These individual cables carry video, audio, intercom signals, and other signals between the camera and the CCU, which is located in either the studio control room or the remote truck. Multicore cable is also used between a camera and a portable VCR. As mentioned earlier, multicore cable is not suited for extremely long cable runs. The maximum distance for multicore is about 2000 feet, while triax can be run a mile or more. Multicore cable permits video return to the camera viewfinder, along with many other functions, such as

- audio from camera to VCR
- diagnostics from VCR to camera viewfinder
- VCR pause/start from camera
- intercom to and from the CCU
- DC power to camera

One final type of cable capable of carrying video signals is quite different from the others. It is **fiber optic** cable. Fiber optic lines are already being used extensively for voice (telephone) and data transmission. Before a video signal is transmitted via fiber optic cable, it is usually digitized by a codec. An uncompressed video signal, once in the digital domain, requires approximately 125 Mb/s for transmission. However, using compression technology, audio and video signals can be efficiently carried by fiber. With its promising capabilities, many consider fiber optics to be the choice of the future.

COAX: Short for *coaxial.* A cable composed of a single conductor core surrounded by a braided shield. Coax is the type of cable typically used to route video signals from one component to another. RG-59 is a common grade of coax video cable.

TRIAX: Short for *triaxial video cable.* Used in situations that require remote video control of cameras at great distances (i.e., multicamera sporting events). Triax allows multiplexed signals to be transmitted great distances over a fairly small diameter cable.

FIBER OPTICS: A technology based on the conversion of electronic signals to digitized pulses of light that can be transmitted through bundled glass fibers. Some predict that fiber optics will someday replace coax, microwave, and satellite transmission technology.

Video Connectors

Fortunately, fewer connectors are commonly used for video than for audio. By far the most common professional video connector is the **British Nut Connector (BNC) connector**. The BNC connector is used with coaxial cable and uses a locking twist-on/twist-off design to ensure that the cable remains connected even with rugged use. BNC connectors can easily be installed onto coax cable by using a stripping tool to prepare the cable and a crimping tool to install the connector. This is a common task at any station or production house; it seems that additional cables of specific lengths always need to be built.

An older and less common video connector is the PL-259, also known as a **UHF connector**. The UHF connector has a larger diameter than the BNC and has a threaded cap that is screwed onto the receiving jack.

Another type of connector, commonly used with consumer video gear, is identical to the RCA or phono connector used for audio cabling. However, remember to use cables rated for *video* signals when attaching video components with phono connectors. Audio cable with phono connectors may not be able to pass the higher frequencies required by the video signal.

The last video connector is used for an RF-modulated video/audio signal. The **F-connector** is a small threaded cap with a single conductor wire protruding from its center. This type of cable and connector is installed by the cable company when it wires a home for cable television.

Adaptors are available to connect or convert BNC to UHF or phono. They can be useful for remote productions where unusual situations and configurations may be encountered.

Self-Study

Questions

1. The most common video connector in a broadcast installation is the:
   a. F-connector
   b. BNC
   c. phono

2. An FCC-mandated broadcast bandwidth of 4.2 MHz limits the actual horizontal resolution of broadcast video to:
   a. 525 lines
   b. 450 lines
   c. 340 lines

3. Saticons, Plumbicons, and CCDs are all types of:
   a. video imaging devices
   b. video monitors
   c. video pickup tubes

4. Standard video cable, such as Belden RG-59, that has a metallic single core surrounded by a braided shield, is known as:
   a. fiber optic
   b. triax
   c. coax

5. A legal broadcast video signal (picture information only) can range from 7.5 to 100 IRE (+4 IRE), providing a contrast range of approximately:
   a. 10:1
   b. 30:1
   c. 100:1

6. The gain switch on a video camera:
   a. boosts the output of the video processing amp
   b. increases the noise in the signal
   c. both a and b

7. The luminance signal is derived primarily from this tube in a three-tube color camera:
   a. red
   b. green
   c. blue

8. One function of the CCU is to:
   a. transcode composite video signals into RGB components
   b. provide remote control of camera functions, e.g., iris, pedestal, and shading
   c. provide test patterns for camera setup

9. The letter Y is used to represent this part of the video signal:
   a. luminance
   b. yellow
   c. subcarrier

10. NTSC video specifies a frame rate of _____ per second, a field rate of _____ per second, and _____ scanning lines.
    a. 2, 24, and 525
    b. 30, 60, and 525
    c. 2, 60, and 340

Answers

1. a. No. F-connectors are commonly used when dealing with an RF signal. The antenna or cable television signal feeding a TV receiver most likely uses an F-connector.
   b. Yes. The BNC connector is a twist-locking connector used on the ends of coaxial cable.
   c. No. However, phono (or RCA) connectors are sometimes used as video connectors for consumer video equipment.

2. a. No. Although the NTSC signal starts with 525 lines, the resolution of the broadcast signal is actually considerably less.
   b. No.
   c. Yes. The formula for determining horizontal resolution is approximately 80 horizontal lines of resolution for each MHz of bandwidth.

3. a. Yes. All three convert light into an electronic signal for the purpose of creating a video image.
   b. No. Like monitors, these devices are indeed transducers but at the other end of the signal flow.
   c. No. Although both Saticons and Plumbicons can be considered camera tubes, the CCD is definitely not a tube.

4. a. No. Fiber optic cable is actually a super-thin glass strand and has no metallic properties.
   b. No. The fact that the cable has two conductors should be a clue.
   c. Yes. The term coaxial describes the construction of the cable; it is made up of a central conductor surrounded by an insulating layer, which is surrounded by an outside conductor, usually braided.

5. a. No. However, the contrast range experienced by many viewers does fall to 10:1 or even lower due to the poor lighting and otherwise compromised viewing environment in the home.
   b. Yes. However, the actual contrast ratio will be limited by the weakest link in the chain. That may be the camera, tape machine, or monitor.
   c. No. Currently, only film can approach this range of contrast.

6. a. True, but b is also correct, making c the best answer.

   b. True, but a is also correct, making c the best answer.

   c. Yes, both a and b are correct. Answer a describes the benefit of using the gain switch, and b explains the negative aspect of the same.

7. a. No. Only 30% of the luminance signal is derived from the red tube.

   b. Yes. The luminance signal is made up of 30% red, 59% green, and 11% blue signals.

   c. No. In fact, the blue signal is the least significant component, making up only 11% of the luminance signal.

8. a. On the contrary, the CCU actually encodes RGB signals to produce a composite video signal from the camera.

   b. Yes. This allows the video engineer to make these adjustments for all cameras from one central location.

   c. No. Camera test patterns are typically placed in front of the camera lens.

9. a. Yes. The letter Y stands for luminance, e.g., R-Y, B-Y, or Y, I, Q.

   b. No. Yellow, by the way, is achieved by combining the primary colors red and green.

   c. No. Subcarrier is sometimes designated by the abbreviation SC.

10. a. No. However, one of the numbers is correct.

   b. Yes. Although if you want to be nit-picky, the frame rate is 29.97, and the field rate is 59.94 per second.

   c. No. You may have confused these numbers with others given in this chapter.

Projects

Project 1

Have a video engineer at your facility show you the inside of a three-tube or three-chip color television camera.

Purpose

To visualize the components of a color television camera and to become familiar with their location within the camera head.

Advice, Cautions, and Background

1. Only a qualified engineer should open up a television camera. There are too many adjustments, switches, and delicate components within the camera head for inexperienced fingers.

2. Due to miniaturization and compact camera-head design, some components will be difficult or impossible to see.

Project 2

Look at the output of your camera on the best monitor available. Compare the pictures with the gain switch at 0 dB, at 9 dB, and 18 dB.

Purpose

To observe the increase in video noise created by the gain circuitry.

Advice, Cautions, and Background

1. Be sure to shoot a scene that has some black areas; video noise shows up quickly in the dark areas of the picture.

2. Make sure the camera is on automatic iris. Notice how the iris stops down to compensate for the increased gain. Remember, each 9 dB of gain should equal 1 1/2 f/stop.

Chapter 4
TESTS and MEASUREMENTS

Anyone who has been in a control room or editing suite and has seen two color monitors side by side knows that color monitors seldom provide an absolute reference on which to base critical viewing decisions. In fact, edit suites often avoid having two color monitors displaying the same picture side by side because the client will want to know why the picture looks different on the *other* monitor and which one is the *true* picture. Even with the best color monitors, what you see is not always what you get. Monitor adjustments, as with adjustments of most television equipment, tend to drift and need constant attention to maintain strict operating tolerances. To provide an absolute reference and to assist in the maintenance and adjustment of this equipment, facilities use test equipment. More specifically, they use specialized oscilloscopes, test pattern generators, and reference camera charts. The specialized oscilloscopes, the **waveform monitor** and the **vectorscope,** are the focus of this chapter.

Both the waveform monitor and vectorscope are dedicated oscilloscopes designed to display the video signal as a linear trace. (An oscilloscope is a device for charting an electronic signal against time on a cathode ray tube [CRT].) The waveform monitor provides a visual interpretation of the video signal level. Although it includes information about the chrominance portion of the video signal, it deals primarily with the video luminance information. The vectorscope is used principally to provide a visual display of chrominance information. It should be noted that a number of test equipment manufacturers have waveform monitor/vectorscope combination models. These scopes combine the features of the waveform monitor and the vectorscope into one unit. This typically saves a little on price and half of the rack space required by two separate units.

Both the waveform monitor and vectorscope are critical to the proper maintenance and adjustment of video equipment. Most importantly, they provide a reference less subjective than the human eye when determining proper levels. Although they can be intimidating at first, the waveform monitor and vectorscope are not just for engineers. At the very least, everyone who works in video production should know what a good signal looks like when it is displayed on a waveform monitor and a vectorscope and what problems look like as well. That way, you will know when you need to call an engineer to take a more exacting look.

Neither the waveform monitor nor the vectorscope change or modify the video signal in any way. Rather, they provide a visual display of the signal so that other video devices, e.g., CCUs, proc amps, and time base correctors can be adjusted or tweaked with the confidence that the adjustments are based on measurable, objective values rather than mere subjective visual observation. To be sure, the human eye can sometimes outperform test monitors. However, for day-to-day operations that require testing and adjustment of the video signal, the waveform monitor and vectorscope are indispensable.

The Waveform Monitor

The waveform monitor is designed to display the video signal at its standard level of 1 volt peak-to-peak across a 75-ohm load. The waveform monitor's graticule (the ruled transparent faceplate through which the linear trace is viewed) is divided into units, from −40 to 120 IRE on one side and from 100 to 0% on the other. The 100 to 0% scale refers to modulation amplitude and is used primarily by the RF engineer. For production

WAVEFORM MONITOR: A dedicated oscilloscope used to monitor and evaluate the video signal. A trace of the video signal is displayed on a scale of −40 to 100 IRE units. Video levels, timing information, and blanking signals can be monitored by using a waveform monitor. The waveform monitor and vectorscope provide an objective reference when setting levels or maintaining video equipment.

VECTORSCOPE: A dedicated oscilloscope used to monitor the color information of a video signal. Hue and saturation are displayed on a circular grid corresponding to the phase and amplitude of the signal's chrominance information. Usually used in tandem with a waveform monitor.

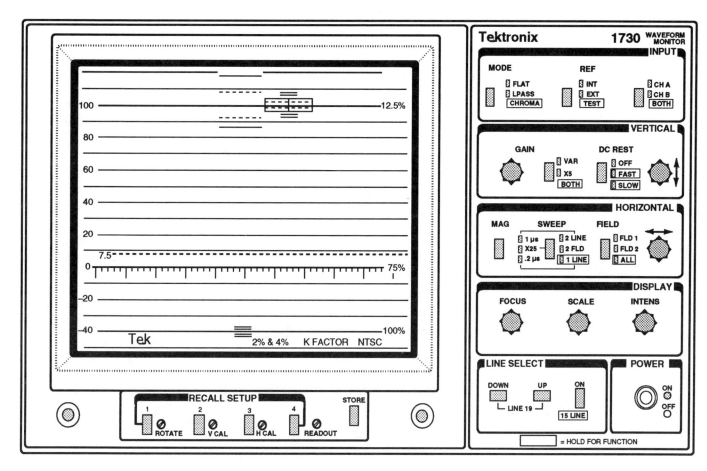

Waveform Monitor

(Copyright 1991 Tektronix, Inc. Adapted with permission.)

personnel, the IRE scale is much more useful. This scale translates voltage into **IRE** or **IEEE units**. (IRE is the abbreviation for the Institute of Radio Engineers, which has been renamed the Institute of Electrical and Electronic Engineers, [pronounced "Eye triple E"]. Although for a time the IRE unit was renamed the *IEEE unit*, it is now known again by its original name.) The IRE unit is the standard unit of measure of the amplitude of the video signal. From blanking (0 IRE) to white (100 IRE), the video signal is divided into 100 equal units. Including sync, which extends from 0 to −40 IRE, the composite video signal is 140 IRE units. The video signal is 1 volt, thus 1 volt equals 140 IRE, and 100 IRE equals 0.714 volts. For those who are more interested in the visual image than its electrical value, it may be helpful to understand that every 20 IRE units is equal to approximately one f/stop. An increase of one f/stop is perceived by most people as a doubling of the brightness of a scene. In summary, opening the iris of the video camera one stop will result in an increase of 20 IRE units on the waveform monitor and a visual perception of a twofold increase in brightness.

IRE: Institute of Radio Engineers. IRE units are used to measure the amplitude of a video signal. On a waveform monitor, blanking is at 0 IRE, white is at 100 IRE, and the tip of sync is at −40 IRE.

IEEE: Institute of Electrical and Electronic Engineers, formerly the Institute of Radio Engineers. The scale on a waveform monitor is divided into IRE, or IEEE, units.

Before using a waveform monitor or vectorscope, it is important that it be properly calibrated. Most scopes generate an internal signal for calibration. A properly calibrated waveform monitor will indicate a full one volt signal peak-to-peak from –40 to 100 IRE units. If the signal deviates by more than 2 IRE units, check to be sure the gain control is set to the calibrate mode. Also, make sure that the waveform monitor is terminated properly. Caution: Do not confuse the calibration controls (normally screwdriver adjustments) with variable gain controls. Once calibration has been completed, the controls for focus, display brightness, and position can be adjusted for optimum reading.

Operating Levels

The video signal, using color bars as an example, should extend from –40 to 100 IRE units as measured from the tip of the sync pulse to the tip of the brightest video, in this case the 100% white bar. The 75% white (actually light grey) bar that is part of the color bar signal calculates at 77 IRE units when the 7.5 IRE setup is considered. Society of

---

NORMAL VIDEO LEVELS

–40 to 0 IRE = sync
–20 to+20 IRE = color burst
0 IRE = blanking
7.5 IRE = black level,
pedestal, setup
77 IRE = 75% white bar on
color bars
100 IRE = 100% white,
clipping level

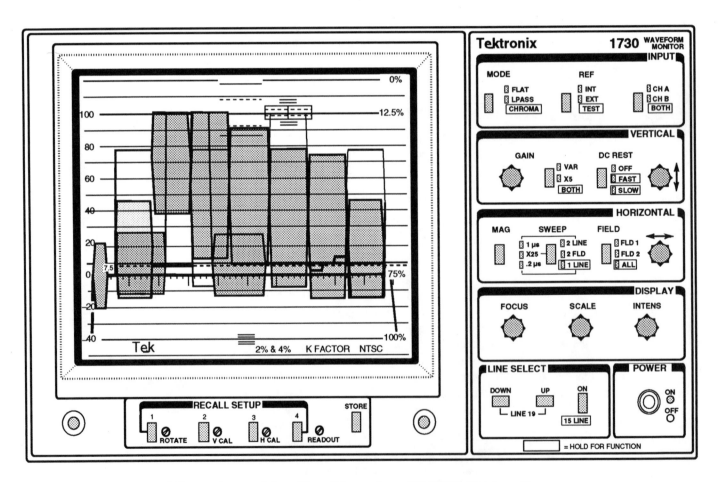

Waveform Monitor Showing SMPTE Color Bars

(Copyright 1991 Tektronix, Inc. Adapted with permission.)

Motion Picture and Television Engineers (SMPTE)-approved and split-field color bars include a 100% white bar in the lower third of the screen in addition to the 77% white bar. The video portion of the picture (as viewed using the IRE or luminance-only mode) should extend from 7.5 to 100 IRE, and only sync information should extend from 0 to −40 IRE. Looking at a waveform monitor display of a video signal, you might accurately surmise that the portion of the display near the bottom of the video picture scale (between 7.5 and 100 IRE) represents black to dark grey values and the display near the top of that same range represents light grey to white values.

Reference black, also known as *setup* or **pedestal**, is usually set at 7.5 IRE. Most waveform monitor graticules have a line labeled *7.5 IRE* to make the adjustment of pedestal easier. The reason pedestal is set at 7.5 units above absolute black, or blanking, is to protect the sync and blanking information from the picture information. This is not nearly as critical as it once was. Now, it is not uncommon to compress blacks much closer to 0 IRE to achieve a wider contrast range or for keying. Raising the black level above 7.5 IRE lowers the contrast range and introduces grain or texture in the black areas of the picture.

White levels are usually set to peak at 100 IRE units. Be careful to read the waveform monitor in the IRE or L-pass mode while measuring luminance information only because chrominance information may legally exceed 100 IRE units and may go as low as −20 IRE units. If the peaks appear to have flat tops, it could indicate that the camera or recorder circuitry is clipping or compressing the whites. Clipped whites should be avoided. However, a high-quality camera may have white compression, which actually provides better pictures in high-contrast, brightly lit situations. The FCC is pretty particular about broadcast transmissions exceeding 104 units because video signals above this level can cause interference with the audio portion of the program. This interference, known as *incidental carrier phase modulation* (ICPM), produces an audible buzz when levels exceed a certain limit. You may have noticed this disturbance when character generator information is inserted into program material or during the broadcast of a commercial spot. If the luminance level of the characters exceeds these limits, it causes an audio buzz and the hot video appears to tear or break up. Although video that is too hot is a problem, the other extreme is equally undesirable. When white levels do not exceed 50 to 60 IRE, the picture will look dim and will have limited contrast. It could be that the camera iris needs to be opened or the VTR's proc amp needs to be adjusted, or it could simply be that the scene is intended to look dark or dull. The important thing is that once you understand the rules, you can sometimes break them, or at least bend them, for aesthetic purposes. To put it another way, as long as transmission regulations are not violated, there is room for creative interpretation of conventional video level settings. Most of the time, however, a video signal that has blacks at 7.5 IRE and whites peaking at 100 IRE will provide the broadest contrast range and the most pleasing picture.

One use for the waveform monitor is to ensure consistency of video levels from shot to shot during the shooting, taping, or editing process. Because flesh tones provide such a standard frame of reference, it is important to keep the exposure of human skin as consistent as possible. Caucasian skin is typically exposed at 60 to 80 IRE, but consistency is much more important than the actual value. The value of the background, clothing, and foreground elements must also be controlled so that the overall ambience of the scene does not change from shot to shot.

PEDESTAL: Also known as *setup* or *black level*, this is the part of the video signal with the lowest level or IRE value. Pedestal is normally set at 7.5 IRE units above blanking (0 IRE on the waveform monitor) and appears as black on the video monitor.

Switches and Settings

One of the first things that you will notice about the waveform monitor is the confusing array of switches and knobs. Before you decide to change careers, consider what a car's dashboard would look like if you were to see one for the first time. A waveform monitor has only a few switches that you need to understand before you can begin to use it for simple signal monitoring.

Under normal operating conditions, you'll probably choose to view two horizontal lines of video. In the normal 2H viewing mode (two horizontal lines), you will also see the horizontal blanking interval, which is between the two lines of active video. The following illustration shows the location of the sweep controls. Most waveform monitors also allow you to view just one horizontal line. Also, you may expand the display to magnify the information in the horizontal blanking area. This expanded view of the video signal is discussed in more detail later in this chapter.

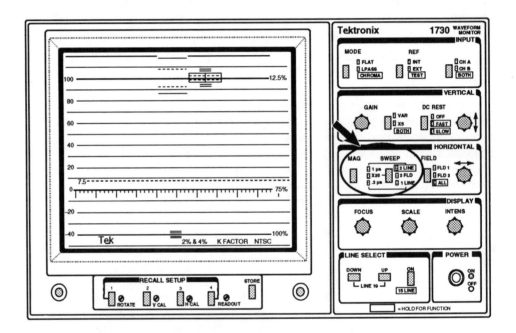

Waveform Monitor's Sweep Controls

(Copyright 1991 Tektronix, Inc. Adapted with permission.)

Another series of buttons allows you to view the filtered video signal in one of three modes: **IRE** (sometimes labeled *luminance* or *low pass*), **flat response**, or **chroma response** (sometimes labeled *3.58* or *high pass*). IRE displays luminance information with almost all of the chrominance information removed; flat response displays both luminance and chrominance information; and chroma displays only chrominance information. The normal procedure is to view the luminance information only (*L Pass* setting), especially if you also have available a vectorscope to monitor chrominance information.

76

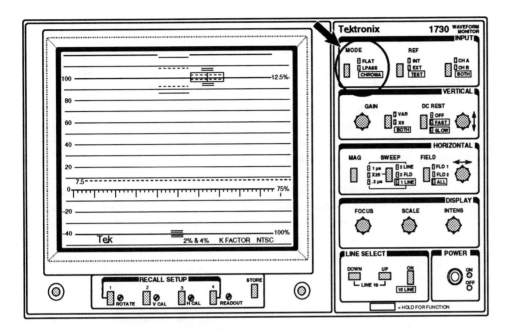

Waveform Monitor's Mode Switch

(Copyright 1991 Tektronix, Inc. Used with permission.)

Almost all waveform monitors and vectorscopes allow the operator to choose from two signal inputs, usually labeled *A* and *B*. A switch allows you to select one or the other. This feature is especially convenient when you have two signals and wish to compare them. When comparing two signals, the source of sync would normally be set to external reference. This facilitates system timing and the comparison of signal levels. Some waveform monitors allow the A and B signals to be superimposed for precise signal matching. In a remote application, you may use the A/B switch simply to allow for dual monitoring of the same signal. The A input may show you the output of the camera, and the B input may be connected to the VTR video confidence out. This allows you to monitor the signal before and after it is recorded by simply selecting either the A or B input.

Another switch found on most waveform monitors is the **DC restoration** switch. When turned on, this switch prevents the display from drifting vertically when the amplitude of the picture information changes. DC restore may be switched off so that the waveform can display problems associated with low-frequency distortion or power line hum. With the DC restore switch in the on position, such distortions are masked.

The waveform monitor will also allow you to choose between **internal** or **external reference** as a source of synchronization information. In most installations, when the switch is set to external, the monitor is referenced to house sync, typically black burst from the sync generator. Black burst is composed of composite sync, reference burst, and a black video signal that is usually at 7.5 IRE above blanking. When the monitor is set to internal sync, it uses the sync from the incoming video as its reference.

Some waveform monitors and vectorscopes offer line select capability. This allows the display of a specific line selected out of all the 525 lines that make up one video frame. One important application is the monitoring of vertical interval test signals (VITS). VITS are test signals transmitted within the vertical interval of the video signal. They allow very precise measurements of signal quality to be taken, even during live transmission.

Finally, the waveform monitor has several knobs and switches that allow you to control the position, focus, and intensity of the display itself. Horizontal and vertical position controls do just that. Remember, however, that there is no "proper" position except when making measurements. For example, the base line or blanking must be on the 0 IRE line when measuring the amplitude of the video signal. On the other hand, you may want to use the horizontal position control to raise the display when measuring the horizontal blanking width. At the same time, you'll probably use the vertical position control to move the leading edge of sync to one of the microsecond divisions on the 0 IRE line. The focus control is used to make the display sharp, and the intensity adjusts the brightness of the display.

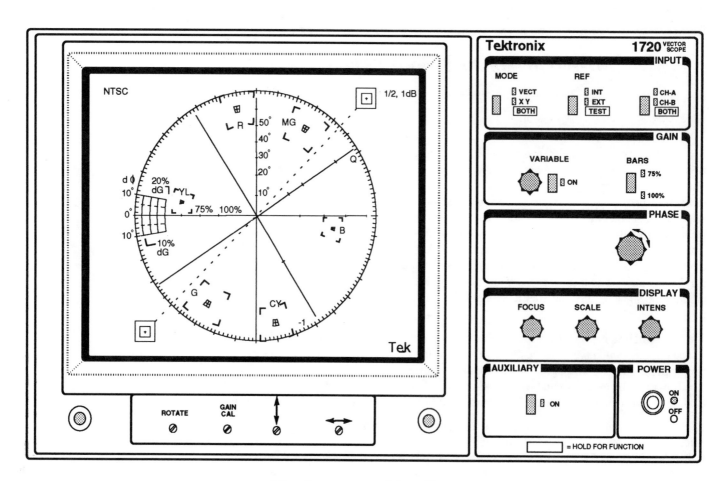

Vectorscope Monitor

(Copyright 1991 Tektronix, Inc. Adapted with permission.)

The Vectorscope

The vectorscope is a dedicated oscilloscope used to monitor only the color, or chroma, information of the video signal. The vectorscope provides a graphic display of both the hue (the actual color) and the saturation (the intensity of the color) of the video signal. The most notable characteristic of the vectorscope is the circular graticule with the location of the various color bar colors forming a ring around the center. The position of the display relative to the **color burst** signal indicates the variations in hue. Hue, because of the way that different colors are derived, is also referred to as **phase**. The color burst signal, which is the reference for all other colors, is a short line protruding from the center toward the left. The vectorscope should be checked for proper calibration before use. When properly calibrated, the color burst will rest at 180 degrees (nine o'clock) and 75%.

The center of the vectorscope indicates the absence of color, or the colors white, black, and grey. Moving out from the center indicates an increase in the amplitude of the signal, which is perceived as an increase in chroma saturation.

COLOR BURST: Composed of nine cycles of subcarrier, this is a reference signal for interpreting the encoded color information. The color burst signal is visible in the horizontal blanking interval on the waveform monitor or on the 180-degree line on the vectorscope.

PHASE: Typically, having to do with the timing relationship of two electrical or electronic signals. In audio, the phase or polarity of two or more signals that are being combined must be the same or low frequency response will suffer. In video, color phase corresponds to hue.

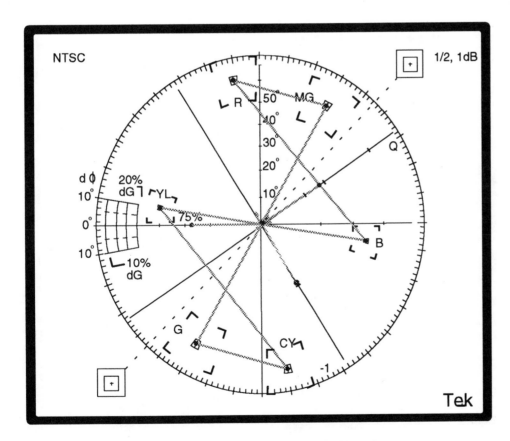

Vectorscope Showing Color Bars

(Copyright 1991 Tektronix, Inc. Adapted with permission.)

79

All vectorscopes have indications on their graticule to represent the FCC tolerances for the colors of the color bars. Each color is indicated by a large box (coarse tolerance) with a smaller box (fine tolerance) inside. When color bars are displayed on the vectorscope, each color should fall as closely as possible within the smaller box. If the vectors fall beyond the boxes, the amplitude of the chrominance signal is too large; if they fall short of the boxes, chrominance levels are too low.

As with the waveform monitor, the vectorscope is used to match cameras, to set up switcher effects (matte keys and background colors), and to set up TBCs for tape playback while viewing color bars at the head of the tape. The vectorscope, used in the external reference mode, is also very handy when performing system timing.

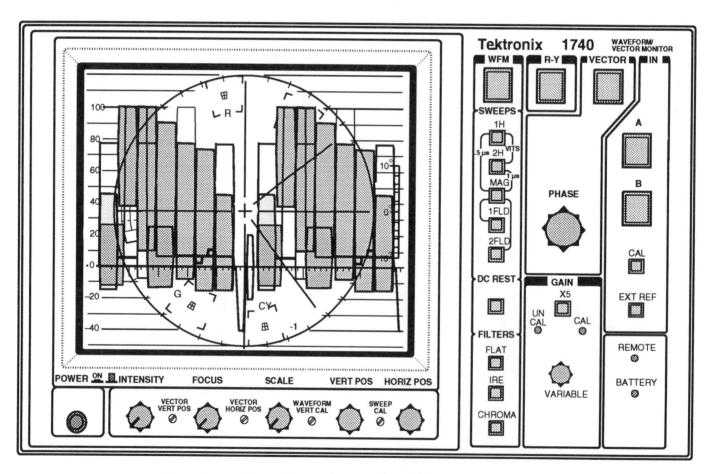

Combination Wavefrom Monitor/Vectorscope

(Copyright 1991 Tektronix, Inc. Adapted with permission.)

Horizontal and Vertical Blanking

As discussed in Chapter 3, the horizontal and vertical blanking interval turn off the electron beam during retrace so that the retrace is not visible. Horizontal blanking takes place at the end of every line (15,750 times a second), and vertical blanking takes place at the end of every field (59.94 times a second). The waveform monitor is essential to monitor the H and V blanking intervals and to ensure that FCC guidelines are being maintained.

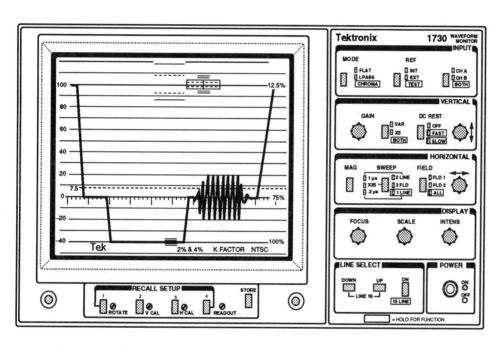

Waveform Monitor Showing Horizontal Blanking

(Copyright 1991 Tektronix, Inc. Adapted with permission.)

It would be helpful to take a look at the horizontal blanking interval displayed on a waveform monitor. You'll have to select 1 microsecond (µs) magnification to be able to see the blanking interval in full-screen size. (One microsecond is equal to one millionth of a second.) The entire horizontal blanking interval should fall between 10.49 and 11.44 microseconds to meet FCC regulations. If the H blanking interval is wider than 11.5 microseconds, it may begin to encroach on the picture area and become visible to viewers at home. It is a good idea to try to keep the H blanking toward the minimum figure because the horizontal blanking will invariably increase at each stage of post-production. To measure the width of the horizontal blanking interval, set the waveform monitor to 2H, flat response, with magnification on. Use the vertical position knob to move the display so that the baseline is at −10 IRE on the graticule. Adjust the horizontal position

a. Horizontal Blanking
   Interval (11.1 μs)
b. Front Porch (1.59 μs)
c. Sync (4.76 μs)
d. Back Porch (4.76 μs)
   d.1 Breezeway (0.56 μs)
   d.2 Color Burst (2.24 μs)

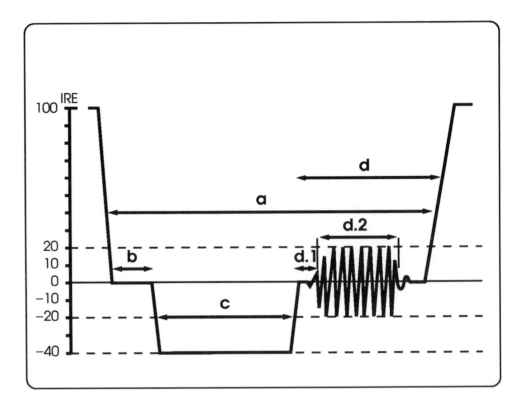

Horizontal Blanking Signal

knob so that the trailing edge of the video crosses the calibrated graticule at the largest tick mark. Count the number of ticks (each one indicates a microsecond) to where the next field's video crosses the line. This is the H blanking width.

The vertical blanking interval is typically made up of 19 to 21 horizontal lines and is sometimes visible on a television receiver when the picture rolls or flips. According to the FCC, the minimum number of lines is 10, and the maximum is 21.5. Within these lines are several signals: lines 1 to 9 contain pre-equalizing pulses, vertical sync pulses, and post-equalizing pulses; lines 10 to 16 have the vertical interval time code (VITC); and lines 17 to 20 carry the VITS. Additional room is available for closed-captioned information.

A video monitor that has pulse cross capabilities (H and V delay switches) can be used to view the horizontal and vertical blanking intervals. This can be especially helpful when adjusting tracking and skew controls for videotape playback.

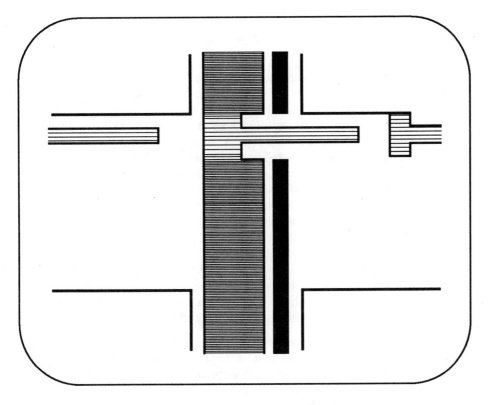

Cross-Pulse Display

System Timing

Color television systems are very demanding when it comes to synchronization. A system must be synchronized horizontally, vertically, and for color phase. Also, it must have the proper color subcarrier to horizontal phase relationship. If these conditions are not met, the result can be picture rolling and tearing or color shifts. Once everything is in synchronization, the next step is to ensure that the signals are arriving in time or in phase with one another. Subcarrier and horizontal timing are essential steps in the installation and routine maintenance of a video system, and the relationship of one to the other, **SC/H phase**, is especially critical when recording on direct, full-bandwidth video recorders. EIA standard RS-170A specifies that whenever two or more signals are combined, usually at a switcher, the **subcarrier (SC)** and horizontal sync pulses must arrive at the switcher in time and in phase with each other. Subcarrier and horizontal sync can vary from one video source to another due to differing cable lengths or the delay differences in different pieces of equipment. A subcarrier mismatch will be visible as a color shift, and a horizontal timing error will be visible as a horizontal jump in picture position. Again, these problems are encountered when one attempts to switch between two live sources or when performing an edit between two tape sources. To compensate for these differences

SC/H PHASE: Relationship of subcarrier to horizontal phase. The industry has standardized SC/H phase so that the subcarrier phase is observed at the leading edge of horizontal sync. On line 10 of color field 1, it should be at 0 crossing and going positive. Some waveform monitors have LED displays to indicate proper SC/H phase.

SUBCARRIER: A continuous sine wave used to encode the color information into the video signal. Subcarrier has a frequency of 3.58 MHz in NTSC video.

in SC and H timing, most cameras and VTRs offer SC and H phase adjustment. Timing a video signal entails making minuscule adjustments because a timing delay of a millionth of a second will cause a visible glitch in the video signal.

Both horizontal and subcarrier phase adjustments should be made using a waveform monitor and vectorscope as a means of referencing the signal being timed to a "prime" source—most often color bars or black from the switcher. Make sure that the waveform monitor and vectorscope are set to external sync when checking the system's timing, otherwise the difference between the two sources will not be displayed. Also, be sure that the switcher output sync processing amplifier has been bypassed. By switching back and forth between the prime source and the signal being timed, the operator can adjust the SC and H phase adjustments at the CCU or the VTR's time base corrector. The waveform monitor set to view the expanded horizontal blanking interval (1μs) is useful to adjust the H timing, and the vectorscope simplifies the adjustment of SC phase. Some newer waveform monitors have built-in SC phase indicators, making the use of a vectorscope unnecessary. When trying to determine the difference in the amount of delay between two video sources, it may be helpful to remember that 1 degree of subcarrier phase is equal to 0.776 nanosecond (ns, one billionth of a second) delay and approximately 6 inches of coaxial cable. (Remember, RG-59, one of the more common types of coax, has a different propagation speed than Belden's 8281, another commonly used coaxial cable.)

It is important to note that the U-matic video recording format, as well as the other heterodyned formats such as VHS and 8mm, do not maintain a locked SC/H phase relationship; instead, the SC and H phase vary continually. To meet FCC requirements, a video signal from one of these formats must be processed through a TBC to restore the SC/H relationship before being broadcast.

75-Ohm Termination

TERMINATION: A 1-volt video signal requires termination by a 75-ohm resistance. Improperly terminated video signals will exceed 1 volt amplitude, and twice-terminated signals will be half of the proper amplitude.

As discussed earlier, a video signal should be exactly 1 volt. To maintain proper signal strength, **termination** of the signal is necessary when routing video signals from one piece of equipment to another. Termination is usually a fairly simple matter. Most video equipment allows the signal to be "looped through," i.e., the signal enters the equipment through an input jack and is available at an output jack to be sent on to another piece of equipment. When a signal is looped through a particular piece of equipment, it should not be terminated at that point. However, termination must take place at the final piece of equipment in the chain. Proper termination requires a load with 75 ohms resistance. This may be provided by an internal termination circuit, or it may be supplied by a 75-ohm resistor cap. If the loop through has a switch located nearby labeled *Hi-Z* and *Terminate*, moving the switch to terminate will provide the proper 75-ohm termination. A video signal that is not terminated will have a 2-volt peak-to-peak level on the waveform monitor. A different problem is encountered when a video signal is terminated twice. In this case, the signal will be only 0.5 volts high on the waveform monitor display. Both an unterminated signal and a double-terminated signal can be a potentially serious problem. On a final note, be aware that looping connections are not the best way to route video signals through the system. One or two loops are usually not a problem, but multiple loops can cause signal loss and reflections. A much better solution is to use distribution amplifiers (DAs) to send a signal to multiple places.

Color Bars

Color bars are a video test signal used widely for system and monitor setup. There are several versions of color bars, but the general idea behind them all is to present a color video signal that can be used as a reference with which to set up video equipment. Once again, it is the waveform monitor and the vectorscope that provide the only objective reference for the evaluation of the color bar signal. The three primary types of color bars are **full-field color bars**, **split-field color bars**, and **SMPTE color bars**. Following, in order, is the standard for color bars of any type: white (actually light grey), yellow, cyan, green, magenta, red, and blue. Some color bars have a 100% white bar and others have a 75% white (grey) bar.

Full-field color bars have colored bars that reach from the top of the frame to the bottom. Split-field bars include a 100% white signal and the –I and Q signals across the bottom of the frame. Testing has proven that humans are much more sensitive to color shifts in the yellow orange (I) region than in other regions. SMPTE color bars are the same as split-field bars with the addition of a strip of reverse bars across the lower third of the frame and a pluge pulse for setting brightness. These additions make the setup of color monitors much easier.

COLOR BARS: A video test signal used as a reference to set up and test video components. Color bars are available in several configurations, including full-field, split-field, and SMPTE standard. The standard array of colors is white, yellow, cyan, green, magenta, red, blue, and black.

SMPTE: Society of Motion Picture and Television Engineers. This organization is actively involved in setting standards for the film and television industries.

SMPTE Color Bars

Audio Monitors

Measurement of the audio portion of the television signal has become more complicated with the increased emphasis on stereo. Audio levels continue to be monitored with VU meters and peak program meters of either the needle or LED variety. However, stereo imaging and phase monitoring require more elaborate electronic equipment. The stereo audio monitor is an oscilloscope like the waveform monitor and vectorscope, but with a different application. The monitor displays the audio signal as what may appear to the novice as a bird's nest of tangled lines. However, to the audio engineer, the stereo audio monitor is a very useful tool for ensuring the proper stereo imaging and mono-compatibility of the audio signal.

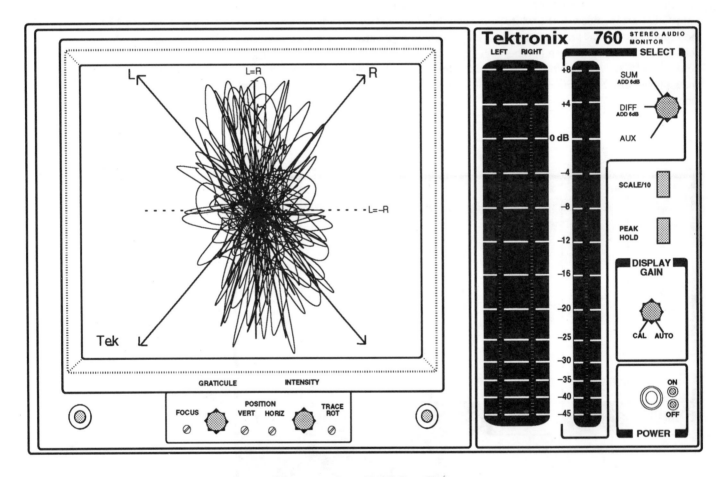

Stereo Audio Monitor

(Copyright 1991 Tektronix, Inc. Adapted with permission.)

Self-Study

Questions

1. The two dedicated oscilloscopes used for monitoring the video signal are the:
   a. CRT monitor and chromascope
   b. waveguide monitor and pulse oscilloscope
   c. waveform monitor and vectorscope

2. The waveform monitor's scale is marked in IRE, or IEEE, units. The video signal from black to optimum luminance should fall within this range, give or take a few units:
   a. 0 to −40
   b. 0 to 100
   c. 7.5 to 100

3. The very center of the vectorscope display indicates the absence of chroma. Such a video signal would be:
   a. saturated
   b. above 100 IRE units
   c. black, white, or some shade of grey

4. In this mode, the waveform monitor strips off the chrominance part of the video signal and displays luminance information only:
   a. 2H
   b. IRE, or L-pass
   c. flat

5. Front porch, sync, and back porch together make up this part of the video signal:
   a. horizontal blanking interval
   b. luminance information
   c. SC/H timing

6. Reference black, also known as setup or pedestal, is normally set at this value on the waveform monitor:
   a. 0 IRE
   b. 5 IRE
   c. 7.5 IRE

7. Video termination requires:
   a. 75-ohm resistance at the end of the video loop
   b. looping inputs to be used in series
   c. that the signal end at a video monitor

8. Color burst, which is visible on the waveform monitor as 8 to 9 cycles within the back porch of horizontal blanking, should be positioned on this line of the vectorscope's display:
a. 0 degrees
b. 90 degrees
c. 180 degrees

9. To perform this timing adjustment, make sure that the leading edge of sync of the source you are timing begins at the same point as your prime source when viewed on a waveform monitor referenced to external sync:
a. horizontal timing
b. vertical timing
c. subcarrier timing

10. This video test signal is usually recorded at the head of every videotape and serves as a reference for setup of the tape on playback:
a. color field
b. cross pulse
c. color bars

Answers

1. a. No. Try again.
b. No. Try again.
c. Yes. Some manufactures have combined both scopes into one device allowing the user to select either monitor with the flick of a switch.

2. a. No. However, 0 to –40 is the value of the video signal from blanking to the tip of sync.
b. No. Black video is usually not allowed to go as low as 0 IRE to protect the blanking information.
c. Yes. Video luminance signals that exceed these values for any length of time can create problems.

3. a. No. Saturation has to do with the presence of chroma.
b. No. It is not possible to determine the luminance level of the video by reading a vectorscope.
c. Yes.

4. a. No. This is the designation for two horizontal lines and is independent of the viewing selection of chrominance or luminance information.
b. Yes. The luminance information is passed through while the other information (chrominance) is filtered.
c. No. The flat setting is used to view both luminance and chrominance.

5. a. Yes.

 b. No. In fact, this part of the video signal is blanked, i.e., no picture information is generated by these signals.

 c. No. Subcarrier and horizontal phase have to do with timing relationships and are not names of video signal parts.

6. a. No. This is the setting for blanking.

 b. No. However, at times it may be set this low, especially when it is the background for a key source.

 c. Yes. It may be interesting to note that PAL and SECAM place black at 0.

7. a. Exactly.

 b. No. However, looping inputs are one place where termination is often required.

 c. No. However, the video monitor is where the video signal is frequently terminated.

8. a. No. Try again.

 b. No. Try again.

 c. Yes. The color burst flag should be at 75% on the 180-degree line.

9. a. Yes. Proper adjustment of the H timing of sources into a switcher will prevent shifts in picture position when performing edits or live switching functions.

 b. No.

 c. No. Subcarrier timing is best performed using a vectorscope.

10. a. No. Color field has to do with the phase relationship of fields 1 to 4 in a composite color video system.

 b. No. Cross pulse, a feature found on certain video monitors, delays the H and V sync so that the blanking is visible on the screen.

 c. Yes. Color bars may be full-field, split-field, or SMPTE standard.

Chapter 5
CAMERA SETUP

REGISTRATION: The process of accurately aligning the output of the red, green, and blue imaging devices to achieve optimum resolution and color clarity of the resulting combined image. Registration involves several adjustment procedures for the red and blue tubes, the most common of which is known as *centering*.

COLOR BALANCE: An adjustment of the output of the red and blue tubes or chips of a color camera to ensure that the camera faithfully reproduces white and all other colors. Color balancing includes white balance and black balance procedures and compensates for various color temperatures of light.

TWEAKER: A small screwdriver (often made of or coated with plastic to prevent shorting delicate circuits) used to make adjustments to video gear. Every good maintenance engineer has a pocketful of tweakers.

Not so many years ago, it took one or more engineers several hours to get a couple of studio cameras up and running and ready for a production. Today, the process not only takes less time but is much easier to perform. New microprocessor-controlled CCD cameras that are set up by computer are much less finicky and much more consistent. Unfortunately, plenty of old tube cameras are still in operation, and it doesn't hurt to know a few things about tweaking them into decent operating condition. For any three-tube color camera, there are two major setup procedures: **registration** (or centering) and **color balance** (also commonly called **white** and **black balance**). Although color balancing is a necessary step in any camera's setup procedure, registration is necessary only with tube cameras. Registration is simply the process of electronically aligning the output of the three tubes so that the individual images created by each tube are perfectly aligned once they are combined and superimposed. If one or two of the tubes are slightly out of alignment, the video image will not be as sharp as it could be. If a tube is severely out of registration, the video picture will have a colored border or outline. The color balance adjustment controls the output of each of the tubes to ensure that colors are reproduced correctly regardless of the color temperature of the lighting source.

Most cameras offer the option of automatic or manual setup. If you have the time and the facilities, manual setup will usually provide better results. The tools and equipment required to perform manual white balance and centering for a typical three-tube camera include a chip chart, a registration chart, a small screwdriver (**tweaking screwdrivers** have plastic ends to prevent shorting delicate electronics on circuit boards), a waveform monitor, and a video monitor. Adjustments can be made at either the camera head (electronic field production [EFP] applications) or at the CCU (studio applications).

The camera charts required to perform some of these setup procedures are available in sets and can be purchased from most video supply companies. A leading name in camera test charts is Porta-Pattern®. The basic charts include the **grey scale** (also known as **chip chart** or **logarithmic reflectance chart**), **registration chart, linearity chart,** and **resolution chart**. Sometimes, a color bar chart and flesh tone chart are included. The grey scale and registration charts are the two that are used to perform the color balance and centering adjustments.

It is important to note that all of the adjustments that will be made to the camera or the CCU involve adjustments to the red and blue tubes only. The green tube is never adjusted when performing centering or white balance because the green tube is the reference tube, and the other tubes are adjusted in relation to the green tube. If the green tube were to be adjusted, the reference would be lost.

Registration or Centering

CENTERING: The rudimentary registration adjustment. Centering is the process of adjusting the H and V position of the red and blue tubes to ensure that the output of all three tubes aligns perfectly.

As already mentioned, registration is the process of aligning the red and blue tubes to make sure that the three separate images, once combined, are precisely superimposed. The several types of registration adjustments include centering, rotation, skew, size, and linearity. Of these, **centering** is the only registration adjustment that should be attempted by anyone other than a qualified engineer.

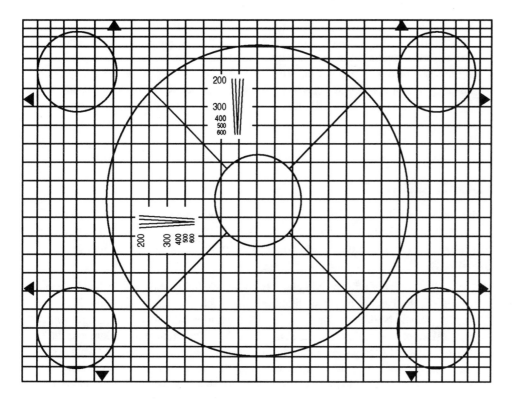

Camera Registration Chart

Following is the basic procedure for adjusting the centering on a three-tube color camera. *Remember, these procedures apply only to a camera having more than one tube. Single tube cameras and chip cameras are exempt from this process.*

1. Frame the registration chart with the camera, making sure that the lens is in sharp focus. This is also a good time to make sure that the **back focus** adjustment is correctly set.

2. Connect the monitor out of the camera or CCU to a monochrome monitor (preferably a large one with high resolution).

3. Using the monitor out switches, select B/–G (this will cause the outputs of the blue gun and a negative image of the green gun to be superimposed).

4. Adjust the H Cent (horizontal centering) and V Cent (vertical centering) adjustment screws for the blue tube so that the two images line up and cancel out each other. Because the output of the blue tube is a black image on a white background and the negative output of the green tube is white on black, when the two signals are aligned, they will combine to produce grey.

5. Now select R/–G and repeat the process, this time adjusting the red tube until its image is aligned with the green tube.

BACK FOCUS: An adjustment to the video camera's lens to ensure sharp focus of the lens in relation to the face of the imaging device. The adjustment ring is typically found on the rear-most part of the lens. Back focus should be adjusted before attempting any registration procedures.

6. You may need to go back and forth from one adjustment to the other until you are satisfied that you have adjusted the tubes as closely as possible. Quite frequently, you will be able to make the images line up perfectly in the center of the frame, but the corners will be out of registration. Your only recourse is to split the difference until other registration procedures can be performed. More advanced registration adjustments are required to correct these errors, and they are best left to a qualified engineer.

Color Balance

Adjusting color balance is in effect adjusting the output gain of each of the three proc amps. Unlike centering, this procedure applies to chip cameras as well as tube cameras. The process is fairly straightforward but must be followed carefully to ensure proper results.

CHIP CHART: Also known as a *grey scale* or *logarithmic reflectance chart*, this camera chart typically has two grey scales, one ascending and the other descending. The chip chart is used when performing a manual color balance on a color camera.

1. Start with an evenly lighted **chip chart** (grey scale), full framed and focused. Be sure to adjust color balance under the lighting with which you will be shooting. Ideally, the lighting should be of one consistent color temperature rather than mixed light sources with different color temperatures. Performing the color balance procedure and then adding another lighting source of a different color temperature will void what you have just accomplished. When color balancing several cameras, it is necessary to position the cameras side by side so that they are shooting the chart from approximately the same angle.

2. Ensure that you have adequate illumination and that the camera iris is correctly set so that the camera is producing a full 1-volt video signal. White peaks should be at the 100 IRE mark on the waveform monitor. Allow the camera to warm up for a few minutes if it is a tube camera. This will allow the electronics to stabilize and help to prevent the white balance from shifting after you've finished tweaking. Chip cameras, on the other hand, require only a few seconds to warm up before you can proceed.

3. Select the proper filter wheel position for the color temperature of the light you are using. Tungsten studio light is standardized at 3200° K, while outdoor sunlight ranges from 5600 to 10,000°K. Most professional cameras provide at least three filter positions in addition to a "blind," or capped, lens position. Once the proper filter wheel position is selected and the iris is properly adjusted, you are ready to begin the white balance adjustment.

4. Set the waveform monitor to 1H and flat response. Looking at the waveform monitor, you should see a series of steps in the shape of a large X. These steps displayed on the waveform monitor correspond to the chips on the chip chart. The highest step, which represents the white chip, should be at 100 IRE, and the lowest step, which represents the black chip (or felt strip) should be at 7.5 IRE. If the black level is above or below 7.5 IRE, adjust the master pedestal control on the CCU until it is positioned correctly. Black levels can also be set with the lens capped, which is a convenient way to achieve absolute black. As mentioned before, the iris control should be set so that the white chip reads 100 IRE. The rest of the steps should be fairly evenly spaced and should cross somewhere between 40 and 60 IRE.

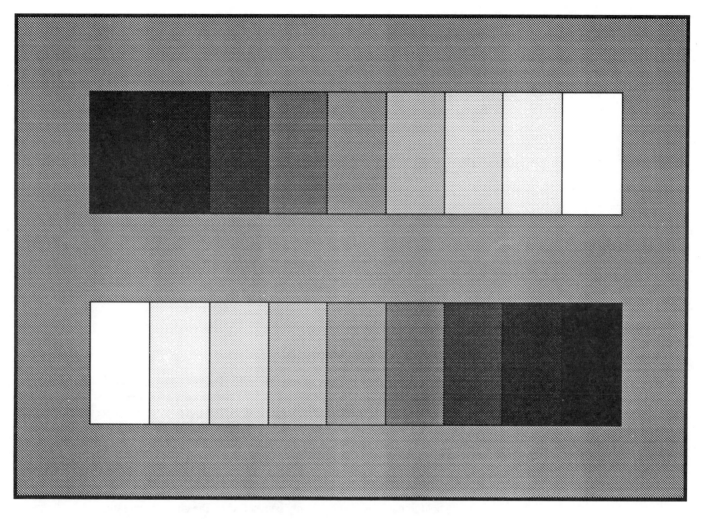

Grey Scale Camera Chip Chart

5. Adjust the gain control of the red and blue tubes until the lines that represent the various steps are as thin as possible. They should change from being fat and fuzzy to thin and defined.

6. Once the black level (pedestal) has been set at 7.5 IRE, the pedestal adjustments for the red and blue tubes can be adjusted to achieve the thinnest possible line at the 7.5 IRE level. It may be necessary to go back and forth between adjusting the gain and ped for each of the tubes to achieve the most accurate white balance.

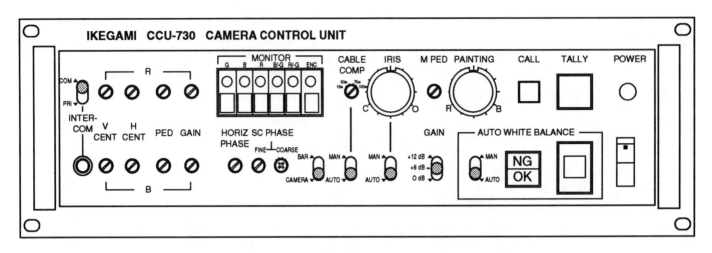

Camera Control Unit (Industrial Grade Camera)

Automatic Color Balance

Almost all video cameras have **automatic white balance** and **automatic black balance** circuits built into the camera head and CCU. The procedure for automatic white balance is to frame a white or neutral grey card (or any other pure, colorless object available, e.g., a white T-shirt), and then flip the switch to tell the camera that this is white. Based on what it "sees," the camera adjusts the output of the red and blue circuits to achieve proper white balance. The automatic black balance procedure is essentially the same, except that you don't need to shoot something black because the camera can achieve absolute black by simply closing its iris during the procedure. Auto white and black balance are normally held in memory and need only be reset when the color temperature of the light source changes.

Creative color balancing can be used to achieve a blue or amber tint to the video if desired. Instead of white balancing on a white card, you may substitute a light blue or pale yellow card instead. Or if you prefer, white balance while shooting through a pale blue or straw colored gel. The camera will correct the output of the tubes and give a colored tint to the video. Color balancing on a blue card will give a warm, amber cast, while balancing on a yellow card will provide a cool, bluish look. The trick is to determine the proper shade of card to use to achieve the desired effect. Be careful to use creative color balancing only in an appropriate and motivated way, or the result may suggest that you failed to color balance properly.

One last adjustment for a video camera is to adjust the camera's viewfinder. In addition to adjusting the brightness and contrast controls to suit your preference, you can also adjust the horizontal hold. If the director constantly asks you to adjust your framing slightly to the left or right, it could be that your viewfinder is slightly off center. Frame an object so that it appears centered for those looking at the control room monitor, and then adjust the viewfinder's horizontal hold so that the object appears centered in the viewfinder.

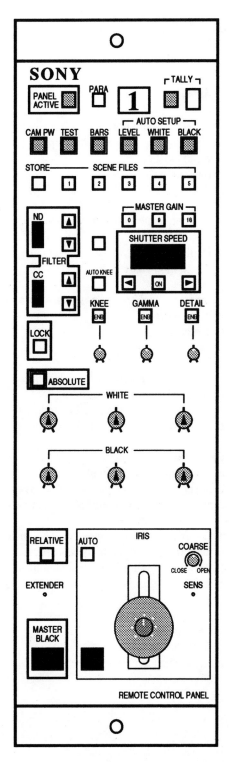

Camera Control Unit (Broadcast Grade Camera)

(Courtesy of Sony Corporation.)

Self-Study

Questions

1. This camera setup procedure is unnecessary when working with CCD cameras due to the fixed configuration of the chips in relation to the prism optics:
   a. white balance
   b. registration
   c. H timing

2. A common name for the grey scale or logarithmic reflectance chart is the:
   a. chip chart
   b. resolution chart
   c. registration chart

3. The primary registration procedure for a three-tube color video camera is:
   a. white balance
   b. centering
   c. back focus

4. The television camera's filter wheel must be reset from 3200 to 5600° K:
   a. before registering the camera
   b. whenever the light intensity changes
   c. when the source of the light changes from quartz tungsten to daylight.

5. The golden rule of registration and white balance is:
   a. never adjust the green tube
   b. never tweak with a plastic screwdriver
   c. always adjust for white balance before registration

6. Before performing a manual color balance, it is important that:
   a. the red and blue channels be switched off
   b. the color monitor be setup and adjusted
   c. the lighting set up be finalized with regard to fixtures, dimming, and gelling

7. This step in the camera setup procedure can be performed with the lens capped:
   a. back focus
   b. white balance
   c. black balance

Answers

1. a. No. White balance or color balance is a necessary step for all professional video cameras and must be performed once the lighting setup has been completed.
b. Yes. Because the chips are permanently cemented to the prism block, the factory alignment of the three chips should never need readjustment.
c. No. Timing procedures are unrelated to imaging technology. However, H and SC timing are important factors in the proper synchronization of signals arriving at a video switcher.

2. a. Yes. The name *chip chart* comes from the practice of using paint chips of the proper shades of grey to make up the chart.
b. No. The resolution chart is used to determine the number of lines of resolution or detail that the camera can reproduce.
c. No. The registration chart is a series of fine horizontal and vertical lines.

3. a. No. Remember, registration has to do with the physical and electronic alignment of the three tubes.
b. Yes. Centering is the process of aligning the output of the three tubes so that they are perfectly superimposed over the others.
c. No. Back focus is a lens adjustment that ensures that the focal point of the lens is in fact focused on the imaging device.

4. a. No. The filter wheel position has nothing to do with the registration process.
b. No. Light intensity has no relationship to color temperature, unless the intensity is reduced by the use of a dimmer.
c. Yes. The function of the filter wheel is to correct incoming light to the color temperature that the camera is set up to expect; in most cases, that is 3200° K.

5. a. Yes. The green tube serves as the point of reference, and the blue and red tubes are adjusted in relation to it.
b. No. In fact, a plastic screwdriver is the preferred tweaking tool.
c. No. The order of these two procedures is not important.

6. a. No.
b. No. Although it's not a bad idea to make sure that the video monitor is properly set up, it should never be used as a reference when adjusting the camera for color balance.
c. Yes. White balance must be performed after lighting is complete, otherwise changes in lighting will affect the color balance of the picture. The only time that this is not true is when you are attempting to use colored lighting for an effect. In this case, the colored lighting should be applied after the white balance has been completed.

7. a. No. Back focus requires that you compare the focal point of the lens throughout its zoom range.
b. No. The white balance procedure depends on the color temperature of the light entering the lens, therefore the lens must be open.
c. Yes. This is because the camera uses the absence of light (a capped lens) as its reference black.

Projects

Project 1

Diagram and label the functions of your camera's CCU.

Purpose

To become familiar with the various controls available on your camera's remote control unit.

Advice, Cautions, and Background

1. If there is a control or feature that you do not understand, consult the operations manual or ask your engineer or instructor.

2. While most CCUs contain many of the same features, the more expensive cameras have additional features. At the same time, however, they may have fewer automatic functions available.

Project 2

Have your facility's engineer demonstrate manual registration and white balance procedures for your studio cameras.

Purpose

To allow you to see firsthand the steps necessary to set up your cameras and prepare them for a production.

Advice, Cautions, and Background

1. Because the CCU and test monitors are rather small, it will be difficult for more than a handful of students to observe the process at any one time. Be prepared to do this in small groups.

2. Once cameras are set up, remember that there is constant tweaking during the course of the production to maintain proper levels and to make sure the cameras match.

Project 3

Perform an automatic white balance of your studio or portable camera.

Purpose

To become familiar with the auto white balance procedure and to see the result on the video output from the camera.

Advice, Cautions, and Background

1. Remember that most cameras retain the most recent white balance settings in memory until the procedure is performed again. Therefore, to see the results of the white balance on a video monitor, you will have to either change the color temperature of the light

source or white balance first on a color that is not white, perhaps a light blue or pink piece of paper. Then, white balance on a white card to see the result.

2. If you can arrange it, the most realistic way to conduct this exercise is to change the lighting from quartz to fluorescent to daylight and white balance under each new light source.

3. Don't forget to set the filter wheel to the appropriate setting before continuing with the auto white balance procedure.

4. Some studio cameras may not have an automatic white balance feature. Some consumer-grade cameras use the auto-white balance designation to indicate continuously self-adjusting white balance circuitry. Avoid using these cameras for this exercise. Ideally, you will have at your disposal a three-tube or three-chip camera, either studio or portable configuration, that has an automatic white balance feature.

How to Do the Project

1. Power up your camera and lighting. Make sure that your lighting is of one consistent color temperature. If you will be using different light sources for different color temperatures, turn on only one source at a time. Set the camera's filter wheel to the appropriate setting.

2. Connect the video out of the camera to a color monitor.

3. Place a white card in front of the camera lens, and fill the frame with the card.

4. Notice the appearance of the card on the color monitor.

5. Activate the automatic white balance circuitry, and observe the effect on the color monitor.

6. Remove the card, and have someone stand in its place so that you can observe the camera's rendition of normal flesh tones.

7. Now change the light source (color temperature), and repeat the exercise.

8. If changing the light source is inconvenient, white balance on a yellow, blue, or pink card, and watch how the camera responds. Observe the effect on flesh tones for each white balance.

Chapter 6
RECORDING

The first commercially available magnetic recording device was the AEG Magneto-phon audio tape recorder unveiled in Berlin in 1935. Twenty-one years later, Ampex introduced the VRX-1000 videotape recorder in Chicago. Since then, the advances in magnetic recording have continued at an astounding rate. Today's magnetic tape and disk recording formats seem to evolve and improve overnight.

Theory of Recording

To understand the developments of magnetic recording, you must begin by looking at the components of magnetic recording—principally the magnetic tape, magnetic heads, and the mechanism designed to transport the former across the latter. The general idea is for the transport device to pull the tape past the record and playback heads of the tape player/recorder. The magnetic heads act as transducers, converting energy from one form into another. In the case of the record head, that means converting the audio signal (electricity) into a magnetic field whose strength is proportional to the signal. The process is reversed for the playback head, which converts the magnetic signal back into an electric signal. It would be best to begin by looking at the dominant recording medium—magnetic tape.

Magnetic Tape

The material on which magnetic recordings have been made has gone through quite an evolution. The earliest attempts at magnetic recording used silk thread in which metal filings and clippings were suspended. Other successful recorders were built using steel wire as the recording medium. For a short time, recorders used steel tape. Unfortunately, this proved to have some serious consequences, not the least of which was its tendency to behave somewhat like a band saw. Further refinements brought about the magnetic tape used today. Constructed of three layers—base, oxide coating, and back coating—magnetic tape has proven to be an effective medium for most recording tasks.

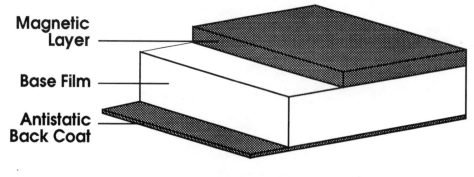

Magnetic Layer

Base Film

Antistatic Back Coat

Magnetic Tape

However, even magnetic tape has gone through extensive changes. In 1928, the base was made of paper. This gave way to cellulose acetate, polyvinyl chloride (PVC), and finally polyester. Polyester, also known as Mylar®, is used almost exclusively today due to its strength and resistance to stretching.

The magnetic property is provided by a flexible coating of metal particles, packed as closely and evenly as possible to achieve optimum performance. The magnetic particles must be aligned along the direction of tape travel and must be adhered permanently to the base. Lubricants are added to reduce wear and increase tape life. To complete the manufacturing process, a back coating is applied to improve the tape's winding characteristics and to reduce the build-up of static electricity.

A primary enemy of magnetic tape and magnetic tape recorders is dirt. Particles the size of smoke can clog heads and prevent proper recording and playback. Video dropouts can result when a particle of dust or dirt as small as 6 microns (1 micron is one-millionth of a meter or 0.00003937 inch) comes between the tape and heads. Dropouts can also be caused by tape that has lost a bit of oxide due to wear. In either case, the dropout on videotape will be visible as a momentary white line across the picture. In particularly dry climates, static electricity builds up and attracts dust and dirt to the tape, which in turn carries the contaminants into the video recording mechanism. To remedy this, tape manufacturers have introduced a line of antistatic videotapes. Scotch identifies them with a red door on their cassettes.

You should keep several things in mind to ensure long life for your videotapes and optimum recording and playback. First of all, tape should be handled carefully. Avoid handling it at all whenever possible. Oils from your hands cause dirt and dust to stick to the tape, which eventually ends up in the machine where they can cause problems. Also, be sure to store the tape in a cool, dry, clean place. According to the National Archives and Records Service in Washington, D.C., ideal storage conditions are 70°F ±10°F with a humidity range of 50% RH ±10%. Cleanliness should be regulated by a filtered ventilation system rated to 50 microns. Videotape should never be spliced. However, if you must splice a tape, do it only for one dub and then remove the tape from service. Ideally, tape should be stored on end to prevent tape edges from damage. Also, keep tape away from strong magnetic fields such as video monitors, electrical generators, and audio speakers. Of course, if you intend to erase the recorded contents of a magnetic tape, bringing the tape into proximity of a strong magnetic field is exactly what is required. This is known as *degaussing* a tape and is performed most efficiently by a commercially manufactured tape degausser. A new tape, one that has never before been recorded, does not need to be degaussed before being put into service. However, it is a good idea to "exercise" a new tape before recording. Simply fast forward the tape and then rewind to the beginning. And finally, make sure that the VCR is properly maintained, cleaned, and in optimum operating condition.

To ensure optimum picture and sound quality, recorders should be cleaned periodically—usually once a day for cassette formats in heavy use, and more frequently for 1-inch machines. Freon TF is a popular cleaner for all metal surfaces in the tape path. However, never use Freon on the rubber pinch roller. Instead, use denatured alcohol. For consumer VCRs, conventional wisdom recommends cleaning only when a problem becomes apparent. Better grade videotape and less intensive use make regular cleaning the exception rather than the rule.

Metal Tape

Metal-particle tape and evaporated metal tape are being used increasingly for newer videotape formats that require that the signal be recorded on smaller amounts of tape surface. Magnetic tape has a wide range of physical properties, depending on its composition. Some tape is easily magnetized and erased, and other tape requires a stronger magnetizing force but retains a stronger signal. These two properties of magnetic tape are **retentivity** (rated in gauss) and **coercivity** (rated in oersteds). Standard ferric oxide tape has a magnetic coercive force of approximately 250 oersted (Oe), and chromium oxide (CrO_2) tape is rated at 500 to 700 Oe. The S-VHS format uses a high-energy oxide tape rated at 900 Oe.

The highest retentivity and coercivity levels are achieved by using metal rather than ferric oxide tape formulas. Metal-particle tape has a coercivity in excess of 1500 Oe, and its retentivity approaches 2500 gauss. The use of smaller (0.25 micron) pure iron particles has made possible this increase in performance. These particles are so small they would spontaneously burn if they were exposed to air, so the outermost layer must be preoxidized to stabilize the tape. Formats that use metal tape include 8mm, Hi-8, ED-Beta, Betacam SP, M-II, and D-2, and the audio recording format rotary digital audio tape (R-Dat). The new consumer format, Hi-8, uses an even newer metal tape technology: evaporated metal tape. This type of metal tape can be extremely thin—a coating of 0.1 microns on a 4-micron base film. That's nearly 20 times thinner than metal particle tape. However, the thinness of evaporated metal tape is also a liability. Lower frequencies, such as chroma information and audio, are best recorded with a thicker coating. Although luminance recording specifications are improved, chrominance and audio specifications can actually suffer with evaporated metal tape.

Tape Consumption and Recording Density

In 1956, 2-inch quadruplex (quad) machines consumed tape at a rate of 30 square inches per second (2-inch-wide tape at a linear tape speed of 15 inches per second). Since that time, improvements in recording technology and magnetic tape have caused the consumption rate to fall dramatically. Today's VHS recorders use only 0.22 square inches of videotape each second in the extended play mode. However, even that figure pales next to the digital audio tape (DAT) format. At 0.15 inches wide and with a linear tape speed of 0.32 inches per second, DAT has a recording density of 114 million bits per square inch, or 1.3 gigabytes per 120-minute cassette! This is currently the highest recording density of any commercially available recording format.

The Transport Mechanism

The mechanism that transports the tape past the heads is designed to provide a constant and stable speed for the magnetic tape. It should also be designed to maximize tape handling, i.e., high-speed shuttling and quick direction changes without damage to the tape. Audio tape recorders are much simpler devices than videotape recorders; however, both use servo-controlled motors to regulate the speed of the source and take-up reels as well as the capstan and pinch roller assembly, when used. The inherent mechanical flaws associated with using belts, clutches, pulleys, and pinch rollers, especially as they become

RETENTIVITY: A measure of the flux density remaining after the external magnetic force has been removed.

COERCIVITY: A measure of the magnetic field strength required to return the material to a state of zero magnetization, measured in oersteds (Oe). A typical range for videotape is 300 to 1500 Oe.

TAPE COSTS	
D-2	= $1.50 per minute
1-inch	= $1.00 per minute
M-II	= $1.00 per minute
Betacam SP	= $1.20 per minute
Betacam	= $0.45 per minute
3/4-inch	= $0.35 per minute
S-VHS	= $0.20 per minute
1/4-inch audio	= $0.25 per minute (at 15 ips)
DAT	= $0.13 per minute
audio cassette	= $0.035 per minute

All figures are approximate.

worn, result in unavoidable instability in videotape recording and playback. For this reason, the video signal played back by a VTR must be processed through a time base corrector to compensate for timing errors.

Recording Fidelity

One factor that must be considered when discussing the fidelity of the recorded signal is the concept of **writing speed**. Writing speed is simply the speed at which the head contacts the medium, in this case magnetic tape. The bandwidth of the analog audio signal is usually no more than 20 kHz. However, due to the extremely large amount of information that makes up each frame of video, a videotape recorder must be capable of writing and reading a signal of approximately 4 MHz, which is nearly 200 times greater than an audio signal. One of the tricks to recording a high-frequency signal is to achieve a high writing speed. This can be achieved either by moving the tape past a fixed head at high speed or by using a rotating head assembly. The former approach is the one taken by most audio tape recorders, and the latter is the approach taken by all videotape formats and by the R-DAT digital audio recording format.

To get an idea of the writing speed necessary for some videotape formats, compare the following:

- One-quarter-inch open-reel ATRs typically use a writing speed of 7.5 inches per second (ips), 15 ips, or 30 ips (with higher fidelity provided by the faster tape speeds).
- One-inch type C has a linear tape speed of 9.61 ips, which delivers a writing speed of over 1000 ips.
- Three-quarter-inch U-format recorders have a linear tape speed of 3.75 ips, resulting in a writing speed of 404 ips.
- Eight-mm video has a tape speed of 0.56 ips and a writing speed of 150 ips.

WRITING SPEED: The speed at which the record/playback heads contact the magnetic medium. In modern videotape recorders, writing speed is achieved by moving the magnetic tape past heads that are mounted in a rotating head drum assembly.

Audio Tape Recording

The AEG Magnetophon was imported into the United States in 1945 and was quickly improved upon by researchers at Crosby Electronics Laboratories, which was named for its founder, Bing Crosby of radio fame. Mr. Crosby began to use modified versions of the Magnetophon to record his ABC radio program, "Philco Radio Time," in early 1947. Later in the same year, Ampex introduced America's first professional audio recorder, the Model 200. The magnetic tape for the machines was produced by the 3M company in St. Paul and later became well known as their Scotch No. 111. These longitudinal recorders, with fixed heads and tape transport systems that drew the tape past the heads at a fixed speed, became the standard for all audio tape recorders to come.

By the early 1950s, most radio networks and stations had converted from shellac records to magnetic tape for their recording standard. In the mid-50s, Les Paul of electric guitar fame approached Ampex to produce a multitrack audio tape recorder. However, it was not until the mid-60s that multitrack recording caught on, in part due to the inspiration of

George Martin and the Beatles. Today, a number of manufacturers produce consumer and professional audio recording devices, including Studer-Revox, MCI, Otari, Philips, Sony, and TEAC/Tascam.

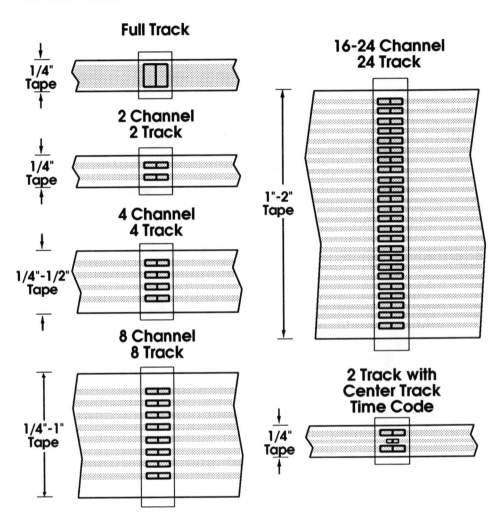

Analog ATR Formats

The Tape Transport System

Most ATRs use a capstan and pinch roller arrangement to move the tape past the heads for recording or playback. The pinch roller is usually rubber, and the capstan is made of a metal or ceramic material. The tape is pinched between the two. The capstan must be perfectly stable and run at a precise speed to provide proper record and playback pitch. Some of the more expensive multitrack ATRs have done away with the pinch roller and capstan arrangement. Instead, they use advanced microprocessor-controlled servo motors attached to the feed and take-up reels to transport the tape.

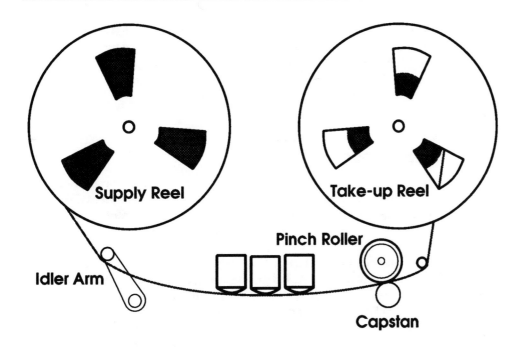

ATR Tape Path Configuration

Except in the case noted where the capstan and pinch roller are omitted, the feed and take-up reels are controlled by less exacting electric servo motors. These must allow the tape to feed smoothly and must take up slack tape after it has passed through the capstan. During fast forward and fast rewind, these motors must handle the tape carefully to avoid stretching or even breaking the tape. To lessen the chance of the tape being damaged during transport, the reels should be as nearly the same size as is possible. The most common sizes are 10.5, 7, and 5 inches; 14-inch reels are used in radio stations with automation.

The tape speed of production ATRs ranges from 1.875 ips to 30 ips or more. The common tape speeds for various ATRs include these:

    1.875 ips = cassettes, some log recorders
    3.75 ips   = consumer reel-to-reel
    7.5 ips   = standard reel-to-reel, cart machines
    15 ips   = high-quality reel-to-reel
    30 ips   = extremely high quality ATRs

Higher tape speed provides several advantages, the most obvious of which is the higher fidelity and greater S/N ratio. In addition, physical or electronic editing is easier because there is additional space between recorded sounds on the tape. On the other hand, slower tape speeds provide greater tape economy, faster access, and smaller reel sizes for the same duration of audio tape.

It should be noted that consumer 8mm, Beta Hi-Fi, and VHS Hi-Fi VCRs make excellent audio mastering decks. Because the audio information is written to the tape by a rotating head, it is capable of extremely high fidelity, even comparable to open-reel 1/4-inch at 30 ips. The Sony PCM system uses a 3/4-inch U-matic VCR as a means of recording digital audio for mastering purposes—but more about that later.

The Magnetic Head

FLUX: The generated magnetic field or lines of force that are produced by the recording head to magnetize the particles in the magnetic tape.

The tape head is simply a coil of wire with a small gap over which the current must jump to complete the circuit. In the record mode, the magnetic field, also known as **flux**, created around the head gap causes the magnetic particles on the recording media to become magnetically aligned. The size of the head gap is inversely proportional to the fidelity of the highest frequencies that the head can record, i.e., the smaller the head gap, the higher the frequencies that can be recorded. Physical limitations prevent the head gap from becoming much smaller than the 0.3-micron head gaps used today. To achieve the high frequency necessary for video recording, faster writing speed is used to realize a smaller apparent head gap. Although most audio tape recorders use a fixed head over which the tape passes, videotape recorders use heads mounted on a rotating drum that spins while contacting the tape. Video recorders and computer disk heads sometimes have head gaps so small that they can become clogged by a particle as small as those found in cigarette smoke. The heads found in audio tape recorders sometimes become clogged by oxide or dirt, in which case they should be cleaned with an approved cleaning solution. Ferrite heads are popular with video recorders; however, they are extremely hard and brittle and may be damaged during the cleaning process if care is not exercised.

Most ATRs have three heads: one to erase the tape, another to record, and a third to play back. They are almost always found in that order (erase, record, and playback), when viewed from the supply reel to the take-up reel.

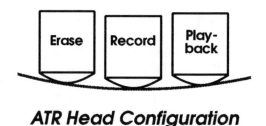

ATR Head Configuration

Sometimes, especially with multitrack ATRs, it is useful to be able to play back selected tracks in synchronization with new material being recorded on other tracks. **Sel-Sync™**, developed by Les Paul, allows the operator to temporarily use the record head as a playback head. Sel-Sync is a trademark of the Ampex Corporation; the feature is labeled by other names on ATRs by other manufacturers. Sel-Sync is necessary in multitrack recording because, at tape speeds of 15 ips or even 7.5 ips, the distance between the playback and record heads causes a delay that makes it difficult to record other tracks in sync.

Bias

Another consideration before selecting tape stock or before beginning to record is the matter of **bias current**. Bias current is an extremely high frequency tone (100,000 to 200,000 Hz) that is applied to the tape by the erase and record heads. The purpose of the bias tone is to prepare the tape to respond to the audio signal. Certain bias tones work better with some magnetic tape stocks than others, so it is necessary for the engineer to set the bias on the machine for optimum response with the brand of tape which the studio has selected, e.g., Ampex 451 or Scotch 806. Following the setting of the bias, no further adjustments to bias are necessary as long as the same tape stock is used.

Basic ATR Maintenance

Aligning the Heads

Five head alignment adjustments are required for most open-reel ATRs. These adjustments should be made and maintained by an engineer or the equivalent. They are zenith, height, tangency, wrap, and azimuth.

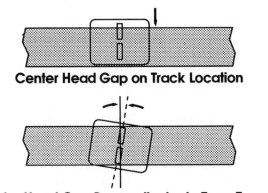

Center Head Gap on Track Location

Make Head Gap Perpendicular to Tape Travel

ATR Head Adjustments

Zenith: adjustment of the vertical angle of the head

Height: adjustment to ensure correct track alignment

Tangency: adjustment to maintain proper pressure of the head against the tape

Wrap: adjustment of the angle at which the tape wraps around the head to ensure optimum contact between tape and head

Azimuth: adjustment to ensure that the head is exactly perpendicular to the tape (reduces the potential for phase problems on multi-track ATRs)

Demagnetizing the Heads

When recordings sound muddy, lacking presence and brilliance, it could be that the heads need to be demagnetized. This is a fairly simple procedure that requires only a head demagnetizer. However, two precautions are in order: First, take care to turn the demagnetizer on and off away from the ATR's heads; and second, be careful not to scratch the heads with the demagnetizer's probe. Some demagnetizer probes are protected by a rubber tip to prevent this from happening. The process is simple; the demagnetizer is switched on, placed close to each of the heads, using a circular motion over the surface of each head, and then pulled away before switching off.

Cleaning the Heads

How often heads need to be cleaned is a matter of some debate; opinions range from every time a new reel of tape is loaded to once a week or so. In any case, it is a fairly simple process. The best way to do the job is to use cotton swabs and denatured alcohol. Clean the heads, tape guides, pinch roller, and capstan—anywhere that the tape contacts the machine. Allow the alcohol to evaporate before proceeding. Be aware that certain head-cleaning solutions may be harmful to any rubber parts, such as the pinch roller. Therefore, if you are using something other than denatured alcohol, read the label carefully.

Videotape Recording

Although television broadcasting and audio recording were both available by the mid-1940s, television recording was not. The first magnetic recording of a video signal was not successfully achieved until nearly a decade later. Before this time, the only way to "record" a television broadcast was to film a kinescope screen. Although this technique allowed rebroadcast of a live television show, the quality left much to be desired. Early attempts at video recording attempted simply to apply existing audio recording technology to the task at hand. As a result, in 1951, engineers at Bing Crosby Enterprises demonstrated a longitudinal video recorder that used 1/2 inch tape and divided the video signal into ten separate channels. Several years later, the BBC went on the air with their own longitudinal video recorders, which they called vision electronic recording apparatus (VERA). The VERA recorder employed a tape speed of 200 ips, which required a reel of tape 5 feet in diameter for a 30-minute program. About the same time, RCA introduced a prototype VTR that used a fixed head and a tape speed of 30 *feet* per second to record 4 minutes of video per reel. It was not until the idea of a moving head mounted on a rotating drum made possible higher writing speed at slower tape speed that the video recorder became a viable device for recording video signals. Finally in 1956, Ampex unveiled the VRX-1000 Mk1 quadruplex VTR, a prototype of the first commercially successful videotape recorder. Using four heads mounted on a head drum spinning at 14,400 revolutions per minute (rpm) to achieve a writing speed of 1560 ips, the VTR could record 45 minutes of monochrome video and audio on a 10-inch reel of 2-inch-wide magnetic tape. The effect on the NAB convention crowd in Chicago was electric.

Videotape Formats

Depending on what criteria you use, there are between 14 and 20 videotape formats available on the market today. The more common formats available for professional and consumer video production include the following:

FORMAT	TYPE	TAPE WIDTH	H LINES	CHROMA
D-1	Digital	19mm	450	4:2:2 (Component)
D-2	Digital	19mm	500	Direct
D-3	Digital	1/2 inch	TBA	Direct
1-inch type B and C	Analog	1 inch	400	Direct
Betacam SP	Analog	1/2 inch	360	Component
M-II	Analog	1/2 inch	360	Component
3/4-inch U-format	Analog	3/4 inch	260	Under
3/4-inch U-format SP	Analog	3/4 inch	340	Under
Betamax	Analog	1/2 inch	240	Azimuth/Under
ED-Beta	Analog	1/2 inch	500	Azimuth/Under
VHS	Analog	1/2 inch	220	Azimuth/Under
S-VHS	Analog	1/2 inch	400	Azimuth/Under
8mm	Analog	8mm	240	Azimuth/Under
Hi-8	Analog	8mm	400	Azimuth/Under

Unlike the 2-inch quad video recording format, all of the formats just listed are **helical scan** recorders, or slant track recorders. The distinguishing characteristic is that they write the information to the tape in a diagonal track across the width of the tape. It would be convenient to be able to assume that the wider the tape used by the format, the higher the quality, but although that was once generally the rule, it is no longer a reality. By using evaporated and metal-particle tape and **azimuth recording** technology, most of the newer tape formats actually record more information onto tape that is narrower and that travels at slower tape speeds than the tape used by the older video recording formats.

HELICAL SCAN: Also called *slant track*, this videotape recording method writes the information to the tape using diagonal or slanted tracks. The name comes from the word *helix*, which describes the fashion in which the tape wraps around the head drum.

AZIMUTH RECORDING: A method by which video information is recorded onto magnetic tape without the use of guard bands between tracks of information. Head gaps positioned at different angles can read closely spaced tracks without interference from neighboring tracks. All of the new, high-quality video recording formats make use of azimuth recording to achieve excellent response from smaller tape surface recording areas.

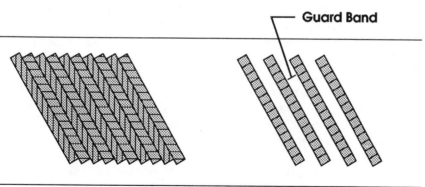

Azimuth Recording   Recording with Guard Band

Aimuth Recording

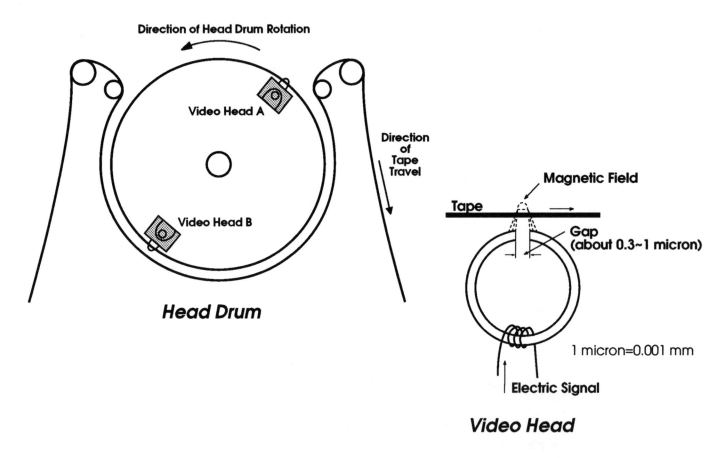

Direction of Head Drum Rotation

Video Head A

Video Head B

Head Drum

Direction of Tape Travel

Magnetic Field

Tape

Gap (about 0.3~1 micron)

1 micron=0.001 mm

Electric Signal

Video Head

Helical Scan VTRs

Helical scan VTRs have several characteristics that must be understood because they affect operation. With the exception of 1-inch type B VTRs, helical scan videotape recorders record a complete field with each pass of the video head. (Segmented recorders, such as 2-inch quad and 1-inch B, record only part of a field with each head pass.) By recording one complete field of video with each head pass, helical scan recorders permit still frames and dynamic motion effects. This definitely is an advantage for helical VTRs—however, that's not the case with the next characteristic. Due to the width of the tape and the tape wrap configuration around the head drum, helical scan VTRs typically have a large tape-to-drum surface area. When moisture condenses on the metal head drum, a condition known as **stiction** can result. Stiction is the adhesion that takes place when two polished surfaces (the head drum and the magnetic tape) come in contact when moisture is present. The result can be that the VTR "eats" the tape, and the damage can be extensive. Most VTRs have a dew sensor that warns of high humidity or moisture on the drum and prevents the machine from operating until the moisture has evaporated. One more consideration is that with a rotating head drum contacting the magnetic tape, leaving a tape machine in the pause mode for too long will wear out both the tape and the heads. For this reason, most VTRs have a long pause feature that automatically slacks the tape after several minutes in the pause mode.

STICTION: The adhesion of two smooth surfaces (in this case, the magnetic tape and the head drum) in the presence of moisture. Most professional VTRs have a dew sensor that indicates excessive moisture in the air (humidity) or on the head drum. The VTR's circuitry prohibits the VTR from being operated until the conditions improve.

114

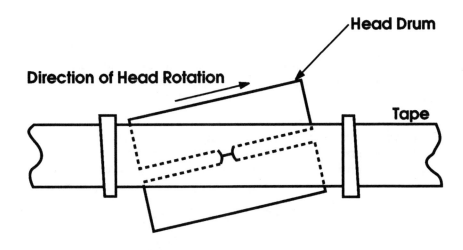

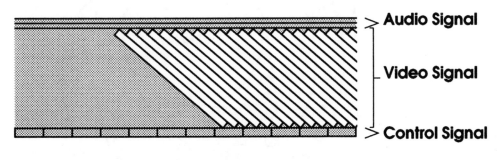

Helical Scan

Variable Motion and Special Effects

As mentioned earlier, one of the advantages of helical scan VTRs over their transverse scan ancestors is the ability to do variable speed and other motion effects. Because the heads write a complete field on each pass, helical VTRs equipped with Automatic Scan Tracking (AST), or **dynamic tracking**, can play back a still frame or slow motion in either forward or reverse. These VTRs make use of special dithering heads to play back a stable signal at speeds typically ranging from −1 to +3 times normal play speed. The head actually flexes to maintain proper tracking at tape speeds other than normal play speed.

A special VTR capable of recording at high speed and playing back at a slower speed was introduced by Sony as Super Slo-Mo for the 1984 Summer Olympics. By recording at higher speeds, the Super Slo-Mo system produces clearer images with less image smear. Similar results can be obtained by using a shuttered video camera to shoot fast motion.

DYNAMIC TRACKING: Also known as automatic scan tracking. This feature permits stable playback at other than standard speeds. Most professional VTRs that have the dynamic tracking feature can play back in variable speeds ranging from −1 to +3 times normal speed.

115

Videocassette Recorders (VCRs)

All of the professional and consumer formats used today, with the exception of 1-inch and 2-inch, are VCRs. They use magnetic tape housed in a plastic cassette rather than mounted on a reel. Cassettes make tape handling easier and faster, and they provide protection from dust and dirt. In addition, the cassette-based formats make playback automation a reality.

All videotape cassettes are self-threading, i.e., once the cassette has been loaded into the VCR, the machine must first load the tape by pulling it out of the cassette and wrapping it around the head drum. Only then is it ready to be played or shuttled. Once the stop button

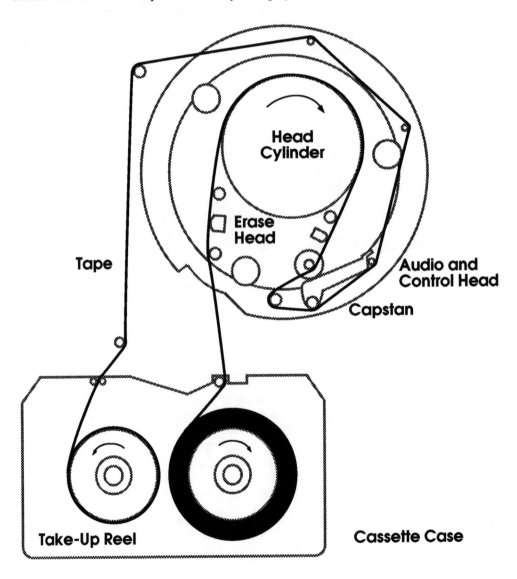

U-Matic Tape Loop

is depressed, the tape must be returned to the cassette before it can be ejected from the machine. This process takes a few seconds but of course is much faster than the manual threading required by reel-to-reel formats.

The 3/4-inch Video Format

The official designation for this format, according to SMPTE, is type E, but 3/4-inch or U-matic (Sony's trade name) is the popular designation. The 3/4-inch format was the first to successfully make use of the color-under recording process, although it has been followed by several commercial successes in the consumer arena, notably VHS, Betamax, and 8mm. In fact, the introduction of the 3/4-inch recorder in the early 70s was initially targeted at the consumer market. Reluctance by consumers was soon offset by interest from educational and corporate video users. Before long, 3/4-inch video was being used for all sorts of applications, including portable recording for broadcast news. However, introduction of the simple-to-use, cassette-based format did create some controversy between engineering and production personnel and management. As with any technological advance, the fear of technical jobs being lost was evident. The reality was that in many instances an engineer was no longer necessary for certain videotape recording or playback situations. Now, a producer could quickly learn the few basic skills necessary to perform the simple operation of a video recording device.

The 3/4-inch recording format is not without its drawbacks. To be able to record the information on tape that was considerably narrower and slower than the 2-inch format required that the engineers design the color-under recording method mentioned earlier. This technique reduces the color information from 3.58 MHz to 688 kHz and records it under the luminance information. The encoding process causes interference between these two signals, which becomes increasingly visible as the number of **generations** increase. The color-under, or heterodyne, recording process is used not only with the 3/4-inch format but also with the consumer 1/2-inch and 8mm formats.

GENERATION: A copy or replication of an audio or video signal. Each generation or copy is further removed from the original signal. In the analog realm, each generation introduces degradation of the signal quality; in theory, digital recording permits each generation to be exactly the same as its source.

Due to competition from the improved consumer formats on the low end and the professional camcorder formats on the high end, the 3/4-inch format's days are numbered. In 1989, Sony sold its one millionth U-matic machine, but just one year later, JVC announced its plans to discontinue its line of 3/4-inch VTRs. That leaves Sony as the only worldwide manufacturer and supplier of U-matic VTRs. Many industry analysts believe that although the U-format's days are short, the large existing base of 3/4-inch users will delay its demise. It is likely to be a few years before the format disappears from vendors' shelves and even more years before the last machines are removed from service.

Super-VHS (S-VHS)

This new format, released in 1986, is notable because although it is being marketed first as a consumer and industrial format, its luminance resolution surpasses that of 3/4-inch SP. This improvement in resolution, coupled with the lower cost of equipment and tape stock, has led some to believe that S-VHS will be the death of 3/4-inch video. Recent introductions of full-featured, three-CCD S-VHS camcorders may result in widespread use of this format for low-budget field acquisition. But first, consider a few specifications of this new format—especially as they relate to the 3/4-inch format.

S-VHS uses a pseudo-component video recording process that allows NTSC artifact elimination, i.e., moiré effect and chroma crawl are reduced but only as long as the signal remains in the S-VHS luminance/chrominance (**Y/C**) component realm. For full system integration, users of S-VHS must use Y/C switchers, signal processors, TBCs, monitors, etc. By boosting the FM carrier frequency of the luminance signal from the 3.4- to 4.4-MHz band (1 MHz deviation) for standard VHS to 5.4- to 7.0-MHz (1.6 MHz deviation) for S-VHS, the horizontal resolution of the luminance signal is increased from 240 to nearly 400 lines. Compare that to standard 3/4-inch with 260 lines and 3/4-inch SP with 340 lines. To enable the recording of the higher bandwidth signal, S-VHS uses a special tape, not metal particle tape but higher energy oxide with 900 Oe coercivity. Also, the head gap is approximately 50% smaller than in standard VHS machines. Like the 3/4-inch format, S-VHS uses color-under recording. However, instead of the 688 kHz used by U-format VCRs, S-VHS uses a 629-kHz chrominance signal, the same as standard VHS. In fact, it is this decision to remain compatible with VHS's low-resolution chroma signal that has become a point of contention with some who would like to use S-VHS for professional applications. Another advantage of S-VHS is that it offers much greater tape capacity—120 minutes versus 20 minutes for 3/4-inch. Although S-VHS tapes cannot be played back in standard VHS recorder/players, standard VHS can be recorded and played back on S-VHS machines, but without the benefits of S-VHS.

Other Consumer Recording Formats

VHS, S-VHS, ED-Beta, 8mm, and Hi-8 are making inroads into the professional broadcast arena. CNN's "Newshounds" is a promotional concept that encourages home video enthusiasts to submit exclusive news footage for air. Television coverage of the Persian Gulf crisis was augmented by the use of Hi-8 camcorders, which were wielded by videographers and producers. When "Video 101," a television show produced for the CBS television network, premiered in the fall of 1988, much of the footage was shot with consumer video gear. Several documentaries shot on the 8mm format have aired on network television and PBS stations. ED-Beta has been marketed primarily to the college and professional sports industry, and 8mm and Hi-8 have been targeted at consumer camcorder enthusiasts.

Recording the Color Video Signal

Composite Video Recording

Of the professional video recording formats listed earlier, most record a composite video signal that is made up of the combined luminance and chrominance information. However, two primary methods are used to record this signal onto tape. The first, which records the chrominance at full bandwidth, is called **highband**, or **direct color**. The second reduces the frequency of the color information and records it under the luminance information. This system is called **heterodyne**, or **color-under**, recording. The 2-inch quad and 1-inch type C recording formats use the direct color recording process to

Y/C: Luminance/chrominance. This designation is used for video signals that keep separate the luminance and chrominance information, thus preventing some of the normal NTSC artifacts, e.g., cross-color and cross-luminance. The S-VHS video format is designed to utilize Y/C video signals.

HIGHBAND (DIRECT COLOR): A means of recording a video signal whereby the chrominance information is recorded at full bandwidth.

HETERODYNE: This method of recording a video signal reduces the chrominance information to a lower frequency and narrower bandwidth before recording. The U-Matic format reduces the chroma from 3.58 MHz to 688 kHz, and VHS reduces it to 629 kHz.

achieve high technical quality. However, all 3/4-inch, VHS, Betamax, and 8mm VCRs use the color-under process to achieve reasonable technical quality at a very cost effective price. The problems associated with color-under recording include these:

- fine detail shows up as chroma noise
- significant quality is lost with each generation
- the signal is not stable enough for direct broadcast without a TBC

Component Video Recording

If you've ever seen a video monitor displaying pure RGB video, you know that the potential exists for extremely high quality images with rich color and sharp resolution. However, once the red, green, and blue signals are combined and recorded on a composite video recorder, much of that quality is lost and cannot be recovered. One practical result of years of HDTV research has been the development of an NTSC video recording system that falls somewhere between the compromise of the standard composite system and the brilliance of a pure RGB video signal. It is known as **component video**. Although economics prevent the component video signal from being a viable transmission format for broadcast and cable companies at this time, processing and recording of the video signal in the component realm has made many converts from those who formerly relied on NTSC composite video processing and recording. Component recording is used for the fairly new **component analog video** (CAV) formats that use 1/2-inch tape such as Betacam and Betacam SP (marketed by Sony and Ampex) and the M-II format (marketed by Panasonic and JVC), as well as one of the digital formats, D-1. Instead of compositing the luminance and chrominance information, they are kept separate as three signals—a luminance signal and two color-difference signals. These three signals (Y, B-Y, and R-Y or Y, U, and V) are recorded onto tape as separate parallel tracks. This is achieved on 1/2-inch tape by using high-density azimuth recording and time compression of the chroma information. The result is vastly superior image quality (better than 3/4-inch and approaching 1-inch) without the bandwidth requirements of full RGB. As long as the signal remains in the component realm, normal NTSC artifacts are minimized, resulting in considerably sharper images.

Another advantage of component video is that many processes such as matting (keying) or color correction are much more successful with a system that keeps the color components separate. Chroma keying is much cleaner with an RGB or component signal than with a composite video source. Posting in RGB would allow keys as clean as from a live camera. One drawback of component video is that it requires three separate wires for routing and processing rather than the one wire required for the composite video signal.

COMPONENT: Y, R-Y, B-Y, or Y,U,V. A video signal in which luminance and chrominance information is kept separate rather than being combined as in the composite video signal. Component processing and routing requires three wires to route the signal, and component recording requires the use of separate tracks on the magnetic tape. Betacam and M-II are two popular component video recording formats.

CAV: Component analog video. This term is usually used to refer to the video recording formats that use component processing and recording of an analog video signal on 1/2-inch magnetic tape. CAV formats include Betacam, Betacam SP, and M-II.

Digital Video Recording

The audio recording industry has rushed to embrace digital recording in a big way, but video technology has lagged slightly behind. However, three different digital video recording formats are currently available. D-1, D-2, and D-3 (previously referred to as *DX* or *1/2-inch digital*) have made possible multiple generations of long-form material with minimal signal loss.

Sony is generally credited with the development of the D-1 format, Ampex with D-2 and Panasonic with D-3. Although each format has its supporters and critics, they really are different enough that their application crossover may be minimal. D-1, which appeared in 1986, is currently the best in terms of picture quality. It is the only digital format that is also component. D-2, introduced in 1988, is seen by many as the logical replacement for aging 1-inch type C format VTRs. Performance is comparable, and the one-cable connectivity of composite video means no costly rewiring and no purchasing of component switchers and other peripherals. D-3, the most recent addition announced at the 1990 NAB convention, uses 1/2-inch tape in a smaller cassette, making it perhaps a more logical choice for location production. Studio versions of the D-3 format accept an extra large cassette said to have a capacity of more than 4 hours record time. Of course, most of this is conjecture at this point, and only time will tell if any or all three of the formats survive this competitive decade. (For a more detailed look at the technical capabilities of the three digital video recording formats, see "Digital Videotape Recorders" in Chapter 9.)

Disk-Based Recording Systems

There is quite a bit of interest in disk-based recording systems for several reasons, the most important being the random access of disk-based recording versus the linear nature of tape. Dedicated magnetic disk recorders, such as the Abekas A-64, are extremely popular in high-end post-production facilities, especially those that work in computer graphics. The drawback to disk recording is the short record durations available. Currently, an A-64 equipped with a 2.64-gigabyte disk can record 100 seconds of full-motion digital video. This is adequate for commercial spot post-production work, but further advances in hard disk and optical disk storage will be necessary for widespread application. Panasonic has recently introduced the first commercially available, rewritable optical disk recorder: the LQ-4000. With its one million erase and write cycles and the capability for recording normal- and high-resolution video signals in the S-VHS, NTSC, and RGB formats, the rewritable optical disk recorder provides both high-capacity recording and random access. Some companies are using optical disk recording and playback systems for off-line editing. Others are using personal computers (PCs) and hard disks to record compressed video for the same purpose. The future of disk-based video recording is promising both for the high end (large post-production houses) and the low end (desktop video) of video production.

Other Recorded Signals

All video formats record more than just a video signal on the videotape. Most formats record at least four separate signals. These include the video signal, two separate longitudinal audio channels, and the control track signal. Optional tracks include longitudinal time code and **audio frequency modulation (AFM)** or **pulse code modulation (PCM)** audio tracks.

PCM: Pulse code modulation. A means of digitally coding an audio signal for recording on magnetic tape. Videotape formats that use PCM audio recording include 8mm, Hi-8, and specific models of 1-inch type C.

Control Track

The importance of the control track signal cannot be overstated. Without stable control track, the VTR will be unable to play back a stable video signal. Control track pulses, one for each frame of video, act as sprocket holes do for film. During playback, the control track signal is picked up by a stationary head. This signal is then used as a reference for tape transport control and head switching.

Cue or Address Track

What some manufacturers call the *cue* and others call the *address track* is usually the place where longitudinal time code is recorded. Like the longitudinal audio tracks, the address track is written to and read by a stationary head.

Videotape Audio

As important as audio is to the enjoyment of a videotaped production, it is interesting to note the wide disparity in the audio fidelity reproduced by the various video recording formats. Some formats use standard longitudinal recording, and others use heads mounted on the video head drum. Most formats record an analog audio signal, but a couple of professional formats and one consumer format offer digital audio recording. **AFM** audio (also known as **Hi-Fi audio**) offers a dramatic improvement in audio recording fidelity and is available on several video recording formats, including M-II, Betacam SP, and VHS and Beta Hi-Fi. By recording the audio signal with heads located on the video head drum, Hi-Fi audio specifications far exceed those recorded on the standard longitudinal audio tracks. The formats that offer Hi-Fi audio often do so because of the poor performance of their linear tracks. This is especially true in the case of consumer formats that have very slow tape speeds. The difference in writing speed for the VHS format is considerable—1.31 ips for the stationary head compared to over 229 ips for the rotating Hi-Fi heads. It is very important to understand, however, that the audio recorded on Hi-Fi tracks cannot be edited separate from the video signal. This makes audio post-production a very difficult matter and usually means that the linear tracks must be used for at least some of the audio program.

HI-FI: Also known as *FM* or *AFM* audio, Hi-Fi audio is a feature available on certain consumer and professional video recording formats that provides extremely high fidelity audio recording specifications. One drawback, however, is the fact that the Hi-Fi audio is depth-recorded underneath the video signal and cannot be edited separately from the video.

Similar results to those gained by the use of AFM technology are achieved by the 8mm format using PCM digital audio recording technology. PCM audio is being used in professional videotape formats as well. Thames Television, the largest commercial television company in the United Kingdom, made a large commitment to the M-II format in PAL but only after Panasonic agreed to develop a new model (the 750) that has two tracks of PCM digital audio. Meanwhile, the M-II VCRs marketed in the U.S. continue to offer only longitudinal and Hi-Fi tracks. Sony and Ampex recently announced new versions of their Betacam SP VCRs with two tracks of fully editable PCM audio. Sony has a 1-inch VTR, the BVH-2800, which has two digital PCM audio tracks, two analog audio tracks, and one longitudinal time code channel. The audio specs of the PCM tracks approach CD quality, which makes this VTR popular for mastering music-related video programs.

Although some consumer video recorders offer Dolby B noise reduction, some professional recorders (the U-matic SP, for example) come with Dolby C. This improves the S/N ratio of the recorded audio signals.

FORMAT	TRACKS	FREQUENCY RESPONSE	DYNAMIC RANGE	S/N
D-1	1 linear	100 to 12 kHz	NA	42
	4 PCM	20 to 20 kHz	90 dB	90
D-2	1 linear	100 to 12 kHz	NA	44
	4 PCM	20 to 20 kHz	90 dB	70
D-3	4 PCM	20 to 20 kHz	100 dB	90
1-inch type C	3 linear	50 to 15 kHz	55 dB	56
Betacam SP	2 linear	50 to 15 kHz	55 dB	52, 72*
	2 AFM	20 to 20 kHz	85 dB	52, 72*
M-II	2 linear	50 to 15 kHz	55 dB	56
	2 AFM	20 to 20 kHz	80 dB	56
3/4-inch U-matic	2 linear	50 to 15 kHz	55 dB	48
3/4-inch U-matic SP	2 linear	50 to 15 kHz	55 dB	52, 72*
VHS	2 linear	50 to 12 kHz	45 dB	45
VHS/Beta Hi-Fi	2 AFM	20 to 20 kHz	80 dB	72
S-VHS	2 linear	50 to 12.5 kHz	52 dB	45
(Optional)	2 AFM	20 to 20 kHz	80 dB	72
8mm	1 AFM	20 to 20 kHz	80 dB	70
	2 PCM	30 to 15 kHz	88 dB	70
8mm PCM	12 (audio only)			

Location Monitoring

One of the advantages of video recording is that it permits instantaneous playback of audio and video for verification. Depending on the format, playback may be achieved by video **confidence heads**, which read the video signal immediately after being recorded by the record heads. Some portable formats do not have confidence heads, and others have no playback capability at all. Of course, in this case, you still have the option of ejecting the tape and playing it in another VCR with playback capability. Almost all

CONFIDENCE HEADS: Playback heads that allow monitoring of the recorded signal while the VTR is in the record mode. Confidence heads permit instant verification without the need to interrupt the recording in progress.

---

* with Dolby C

camera and recorder combinations will allow video playback monitoring through the camera's black-and-white viewfinder. To permit playback through the viewfinder, it is necessary to use the multicore camera cable between the camera and the portable VCR.

Time Base Correction

You may wonder what time base error is and why it needs to be corrected? With video, as with many other things in life, timing is everything. Every second of NTSC video contains 30 frames; each frame is made up of two fields and 525 horizontal lines of picture and synchronization information. Vertical and horizontal sync pulses ensure that each field and line begin at the proper time. There's little margin for error. A new line of video is written to the screen every 63.5 microseconds (a microsecond is one-millionth of a second). If one of these lines begins a little late or scans at a faster rate than the others, the video signal will experience problems. Hooking, tearing, rolling, and color shifts are just a few of the problems that can result from improper timing or corrupted sync signals.

In any video production facility, the greatest source of timing errors are the videotape recording systems. The problem with VTRs is that they are mechanical. The lines of video information must be deposited on a moving strip of magnetic tape by rapidly spinning heads. The constantly changing variations in the video signal after being recorded onto videotape and played back are similar to audio wow and flutter. They are caused by inherent mechanical factors present in all VTRs, e.g., friction and temperature variation. Consider the 3/4-inch U-matic format as an example. The U-format VCR has two video heads mounted on the head drum. Each head records or plays back a single field (or 262.5 lines) every 1/60 of a second. This information is written on a diagonal track 6.5 inches in length and 0.03 inches wide. To reproduce or play back this signal, the process must be replicated exactly. The slightest stretching or shrinking of the tape or the slightest misalignment of the head or tape path will cause the image to suffer in recording and play back. Early video recordings were so susceptible to playback problems that the head drum was shipped with the tape so that the same head drum used to record the image could be used for playback. Time base error is difficult to see if you don't know what to look for. A monitor or television set is quite forgiving of timing errors, so in many cases, timing errors will not be apparent when viewing videotape playback. If the image appears to be fine on the video monitor, there may still be hidden problems. The monitor's automatic frequency control (AFC) may be hiding the problem, or the tape may play back fine on one machine and display a problem on another VCR. Another thing to remember is that time base error is cumulative; it doubles each time that the signal is recorded or dubbed. Although most modern VTRs are precision machines capable of incredible accuracy, for critical signal-processing demands, a **time base corrector** (TBC) is essential.

TIME BASE CORRECTOR: A digital video device used to stabilize the output of a VTR so that the signal can be mixed with other sources.

123

The Proc Amp

The FCC is the government agency responsible for all broadcasts, including those of video and audio information. To fulfill the requirement for licensing by the FCC, a local station must monitor its video signal and maintain it within strict tolerances. Sync, blanking, burst phase, and video amplitude must be within tolerable limits. Whenever the source of the broadcast signal is a VTR, there must be a means by which the video levels can be adjusted. To adjust the video signal from a recorded source, a video processing amplifier (**proc amp**) is used. Most TBCs have built-in proc amps for this purpose. Proc amp controls are typically labeled as follows: *video level, chroma phase, chroma gain,* and *pedestal* or *black level.* The TBC usually serves two purposes: it corrects timing errors as they occur, and it allows adjustments to be made to the levels of the video signal.

PROC AMP: Processing amplifier. The device that allows adjustment of the parameters of the video signal, e.g., video level, black level, chroma level, and chroma phase.

TBCs are an integral part of most broadcast VTRs and are available as an outboard item for industrial and consumer VTRs. These outboard TBCs typically cost between $2000 and $8000. Some TBCs include simple digital effects. A TBC is required to enable a 3/4-inch VCR to be played back to air in a broadcast application or to allow a videotape signal to be mixed with another video signal at a switcher for A/B editing. In an A/B roll editing system, each source machine must be routed through its own TBC. This is to ensure that the signals from the source machines arrive at the switcher in sync with each other.

How a Time Base Corrector Works

The TBC is simply a variable delay device. Unstable video from the playback VCR (jitter video) enters the TBC and is compared to stable sync from the system sync generator. The unstable video is delayed a fraction of a second, depending on how much the signal differs from the reference, and is then released line by line at a stable rate. Occasionally, part of a line is missing due to bad tape or dirt between the tape and the heads. In this case, the **dropout compensator** (**DOC**) inserts similar video information from the previous line.

DROPOUT COMPENSATOR: A feature found on many time base correctors that inserts video information where a tape dropout occurs. The inserted video is taken from the preceding line, and although it does not match the missing material exactly, it is better than seeing a black or white streak.

During this process, the proc amps allow the signal to be manipulated. Video and black levels may be adjusted, and chroma gain (saturation) and phase (hue) adjustments can be made. Also, new sync and burst can be added to the video signal. Now, the signal can be mixed or combined with other signals that have been locked to system sync.

The heart of the TBC is the memory, i.e., the amount of video information that can be stored. The first TBCs were analog devices and used analog delay circuits. Today, TBCs are essentially digital devices, using digital storage to provide the necessary delay time. The memory, usually referred to as the **window**, is rated in horizontal lines. Although early TBCs had only a four-line window, today a 16-line memory is fairly standard. This means that the TBC can delay or advance the signal eight horizontal lines to correct all but the worst timing errors. The falling price of digital memory chips (thanks to the booming computer industry) has made possible low-cost TBCs that have full-frame memory capability. They can store a full frame of video for freeze-frame effects.

WINDOW: For TBCs, the number of lines of memory that the TBC can store for correcting timing errors.

124

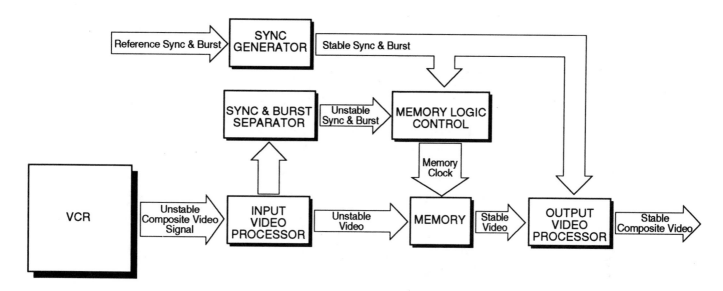

Time Base Correction Flow Chart

A close relative to the TBC is the **frame synchronizer**. This device, which has between one and four fields of memory, can be used to synchronize an external signal coming into the station from an external source when genlocking the source is not possible. This is frequently the case with microwave or satellite feeds. A frame sync will typically retain a freeze frame of the last frame of video if a signal is interrupted. This effect is a common sight on news feeds, particularly during poor transmission conditions.

One note of caution when using a frame synchronizer: Beware of a potential loss of lip sync if the video is delayed without delaying the audio accordingly. Frame syncs, in the process of synchronizing the video with the house system, will delay the video by as much as one frame. As you'll recall, there are 30 frames in each second of NTSC video. Therefore, one frame of video equals 1/30 of a second or 33 milliseconds (ms). When you consider that multiple processing of the video by frame synchronizers will compound the amount of time that the video is trailing the audio, you see how the problem arises. For audio and video to appear in sync, the audio should not lead the video by more than 20 ms nor lag behind by more than 40 ms. Using this definition of lip sync, one pass of video through a frame sync will cause lip sync problems unless the audio is delayed by a comparable amount.

FRAME SYNC: Short for *frame store synchronizer*. This is a digital storage device capable of delaying an incoming, nonsynchronous video signal by as much as one frame to synchronize the signal with house sync.

Enhanced Time Base Correctors

In recent years, digital TBCs have become the standard due to the limitation of analog storage and to the ever increasing cost efficiency of digital storage. Digital storage does not degrade the image and is virtually noise-free. However, the use of digital storage requires that the TBC contain analog-digital (A/D) and digital-analog (D/A) converters.

Once the video image was digitized, it made sense to some TBC manufacturers to add simple digital video effects, such as mosaic, sepia, position, and compression. TBCs with these bells and whistles are aimed at the low-end market whose users might not otherwise be able to afford digital effects.

With the increased interest in component and semicomponent video formats, transcoding TBCs are becoming quite popular. Transcoding is the ability to accept either a composite, component, or S-video (separate luminance and chrominance) input and to provide a similar choice of output signals. Processing within the TBC is performed while keeping the luminance and chrominance signals separate to minimize interference and the accompanying artifacts. Transcoding TBCs are especially popular in an interformat editing system, where you might edit from S-VHS to U-matic or from Hi-8 to Betacam.

Another recent trend is a dual-TBC with effects for use in an A/B roll editing system. Because both source VTRs require a TBC, some manufacturers are putting both TBCs, a video switcher, and digital video effects all in one box. Although this can quickly become limiting, for simple A/B roll editing, the cost effectiveness is obvious.

Self-Study

Questions

1. Betacam SP, M-II, and D-1 are all video recording formats that record the luminance and chrominance signals without combining them. What type of recording is this?
   a. composite
   b. component
   c. segmented

2. Small head gap and this are two factors that permit greater high-frequency response in the magnetic recording/playback process:
   a. high writing speed
   b. wide tape width
   c. balanced inputs

3. Which of the following professional videotape recording formats makes use of the heterodyne, or color-under, recording process?
   a. D-2
   b. 1-inch type C
   c. U-matic

4. When a tape is degaussed, it is:
   a. analyzed for dropouts
   b. erased
   c. recorded with black video

5. Freon and this are the two solutions most commonly used to clean the head of audio and video recorders:
   a. carbon tetrachloride
   b. water
   c. denatured alcohol

6. This Ampex VTR was introduced in 1956 and became the broadcast standard for the next two decades:
   a. 1-inch type C
   b. 3/4-inch U-matic
   c. 2-inch quad

7. The minute variations in tape speed and signal timing that inevitably result from the mechanical nature of all VTRs are known as:
   a. jitter
   b. time base error
   c. both **a** and **b**

8. Which of the following video recording formats is a CAV format?
   a. Betacam SP
   b. D-1
   c. 1-inch type C

9. Many of the newer video recording formats make use of this technology instead of guard bands for the prevention of cross-talk interference between recording tracks:
   a. azimuth recording
   b. color-under recording
   c. companding

10. S-VHS achieves nearly 400 lines of resolution by increasing the FM carrier frequency of the luminance signal, by decreasing the head gap, and by recording on:
    a. metal-particle tape
    b. double-sided tape
    c. high-energy oxide tape

Answers

1. a. No. Composite recording records a signal that is the combined luminance and chrominance information.
   b. Yes. All of these videotape recording formats record luminance and chrominance signals separately on the tape.
   c. No. Segmented recording has to do with recording less than a field of video with each pass of the video head.

2. a. Yes. High writing speed produces greater high frequency response.
   b. No. Today, the width of the tape has little to do with the frequency or complexity of the signal that can be recorded.
   c. No. Balanced or unbalanced inputs have no effect on the recording process.

3. a. No. D-2 is a digital, high-band recording format.
   b. No. One-inch type C is a high-band recording format.
   c. Yes. U-matic and several of the consumer formats use this technique.

4. a. No. However, some tape degaussers will perform this function at the same time.
   b. Yes.
   c. No. However, it may be useful to know that an erased tape may need to be recorded with black before being used as an edit record tape.

5. a. No. Although carbon tet was used at one time, it is generally considered too dangerous for everyday use.
   b. No. Water is almost always an enemy of electronic and electrical components.
   c. Yes. Denatured alcohol is safe for use on both the heads and the rubber pinch roller.

6. a. No. One-inch type C was not introduced until the mid-70s.
b. No. The U-matic format has only had limited acceptance in the broadcast arena, primarily for on-location news gathering.
c. Yes. Now, the 2-inch quad VTR is quickly becoming a museum piece.

7. a. Yes. However, the correct answer is **c**, which includes both **a** and **b**.
b. Yes. However, the correct answer is **c**, which includes both **a** and **b**.
c. Yes. Jitter and time base error are both valid names.

8. a. Yes. Betacam and Betacam SP are component analog video recording formats.
b. No. D-1 is component, but it is a digital component format.
c. No. One-inch type C is analog, but it records a composite rather than a component video signal.

9. a. Yes. Azimuth recording is used in most of the 1/2-inch and 8mm recording formats.
b. No. Color-under recording is used to reduce the recorded frequency of the chrominance information.
c. No. Companding serves other useful functions, most notably improving S/N ratio.

10. a. No. The developers of S-VHS decided against using metal tape to retain close compatibility with the standard VHS format.
b. No. There is no such thing as double-sided magnetic tape for professional audio or video recording.
c. Yes. Magnetic tape with a coercivity rating in the 900-Oe range is required to record the S-VHS signal.

Projects

Project 1

Take a field trip to a television broadcast station or video post-production house where several videotape formats are in use.

Purpose

To observe various VTRs in operation and to compare their advantages and disadvantages and their subsequent function in the production environment.

Advice, Cautions, and Background

1. Locate a production facility that uses several different videotape formats in various applications. Your instructor may arrange the trip for the entire class.

2. Have a videotape operator, editor, or engineer explain the facility's reasons for using one videotape format over another. Ask for a simple demonstration of the machines, e.g., loading, recording, playback, and setup.

3. Be punctual, be prepared with appropriate questions, and be careful that you do not interfere with normal working operations.

4. Prepare a short paper describing the various formats you observed and briefly explain why that format was used in that particular situation.

Project 2

Collect data from broadcast equipment manufacturers, and then compare technical specifications and the cost of the available video recording formats.

Purpose

To understand the many options available and the technical and financial criteria upon which station managers must base their purchasing decisions.

Advice, Cautions, and Background

1. You may choose to contact the manufacturers directly (Ampex, Sony, Panasonic, JVC, BTS, etc.), or you may be successful contacting a local vendor of broadcast equipment. In either case, explain that your interest is educational and not for immediate purchase. They should be quite helpful. If you have a large class, you may choose to have designated individuals contact the vendors or manufacturers so that they are not deluged with calls.

2. Try to acquire information on at least five different videotape recording formats. Be sure to get technical specifications sheets and current price lists.

3. Be aware that equipment manufacturers arrive at technical specifications using nonstandardized testing techniques. Try to keep a healthy skepticism when comparing specifications.

4. Keep in mind that technical considerations are only part of the equation. Price versus performance is often the bottom line. Another very important consideration is compatibility with existing tape footage and with outside suppliers or users of the videotape product.

Chapter 7
POST-PRODUCTION

Video and audio post-production today, more than ever, make up a large portion of the production process, in terms of both time and budget. A smaller and smaller percentage of programs are being produced live, and those live programs commonly have videotaped segments rolled in, segments that were produced using the post-production process. Simply stated, post-production plays a major role in the production process. Although numerous processes are involved in video post-production, the most significant is videotape editing. Reduced to its most basic form, editing involves recording or dubbing one segment of video and/or audio after another until the program or project is finished. Video editing differs from film editing in that it is a transfer process. Instead of cutting and splicing selected shots together, segments are transferred, or dubbed, from one VTR to the other. (For the record and as a historical aside, for several years after the invention of the videotape recorder, videotape editing also was a physical slice-and-splice routine.) As a consequence of the electronic dubbing technique, the edit master videotape is made up of video and audio material that is one generation removed from the original material. However, on a positive note, the source footage remains intact and undisturbed.

Although all of this may sound simple enough, much of the complexity of the video editing process results from the challenge of having to make creative decisions while respecting the technical requirements associated with the recording of complex electronic signals on a moving strip of magnetic tape. Although editing is first and foremost a creative matter, this chapter covers the technique of post-production from the perspective of the technical process. This is not to diminish the creative aspect of editing. Rather, the focus is on the mechanics involved so that you can concentrate on the creative process later without being distracted by the necessary technical details.

One thing that makes video editing different from video recording is the simple matter of accuracy. Editing involves the recording of precise segments in an orderly way using exact in and out points to specify where a visual or sound begins and ends. In fact, the most common activity associated with the editing process is the selection of in and out points for both the source footage and the edit master videotape. Because a computer excels at dealing with numbers, in and out points are typically identified by a numerical value so the computer edit controller can identify them. The assigning of a numerical value to tape positions is achieved by using control track pulses, or time code. (Time code is considered in detail later in this chapter.) Once the computer has identified the number relating to the in and out points of the selected material to be edited, the edit may be performed. Actually, because the computer is adept with numbers, you need only provide three of the four values (source VTR in, source VTR out, record VTR in, and record VTR out), and it will compute the fourth.

Part of the process that the VTRs must perform to execute an edit is to transport the tape to the exact frame location specified before beginning to record. This takes place during what is known as the **preroll**. For clean edits to be performed, the source and record VTRs must roll together—locking up at proper speed to arrive at the selected in point in sync with each other. The computer "bumps" the VTR during the preroll, making subtle speed adjustments so that both VTRs arrive at the edit point at the same time. Most professional VTRs can lock up with as little as 3 seconds of preroll, and most editing systems are programmed to provide a 5-second preroll, although this duration is usually adjustable from 1 to 10 seconds or more. The one major consideration is that for preroll to work, there must be the necessary amount of stable video (preferably 5 to 10 seconds) before the selected in point. Without this stable video, the VTRs may be unable to lock up during the preroll time.

PREROLL: This is the time it takes for a VTR to get up to speed in preparation for an edit. Most professional VTRs can get up to speed in as little as 3 seconds; however, 5 seconds is the normal preset duration for most editing systems.

Insert vs. Assemble Editing

An editing VTR allows the selection of tracks to be recorded, including the video, audio, address, and control tracks. The difference between insert and assemble editing has to do with this ability to select which of these signals will be recorded. Assemble editing records all of the signals at once. Video, audio, sync, and control track signals (if present) are first erased and then recorded with the new signal. With assemble editing, no provi-

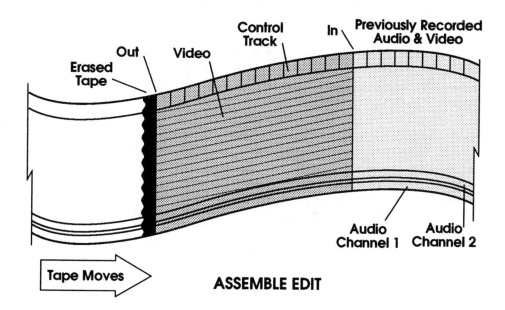

ASSEMBLE EDIT

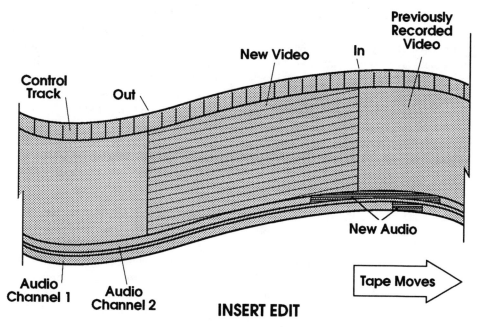

INSERT EDIT

Insert and Assemble Editing

133

sion is made for retaining any information that may have been on the tape previously. As a note of caution, the control track signal recorded in the assemble edit mode is only as stable as the control track from the source footage. If the footage was recorded on location with a portable VCR, there's a reasonably good chance that there will be occasional instability.

Insert editing, on the other hand, permits the operator to select whether video or audio signals will be recorded. One drawback to insert editing may also be considered an advantage—stable control track must already be recorded onto the tape onto which you want to perform insert editing. The process of recording continuous control track on a tape to be used for insert editing is known as **blacking**, or **striping**, a tape. This step requires a little extra time before editing can begin; however, the advantages usually far outweigh this inconvenience. Insert editing onto a tape that has been blacked, or striped, allows you to select video only or any of the available audio channels for recording without affecting the other signals on the tape. Because the control track is continuous and uninterrupted, edits are cleaner and have less instability at edit points. Perhaps the greatest advantage of insert editing over assemble editing is that insert editing does not erase the control track and video and audio signals ahead of the newly recorded signals. Because this is the case with assemble editing, the only way to assemble a program is from beginning to end. Any attempt to lay down an assemble edit in the middle of a series of shots will result in a large glitch at the end of the assemble edit. This glitch is the result of the erased tape that precedes the newly recorded material.

BLACKED TAPE: A tape that has been prepared for insert editing by having stable video and control track recorded for its duration, or at least a bit longer than the length of the program to be edited. The term *blacked* is used because a video black signal is commonly used as the source of video and control track, although any stable video signal will do the job. The term *striped* is often used to refer to a blacked tape that has also been recorded with time code. It is important that the video, control track, and time code signals be recorded so that they are locked and synchronous with each other.

On-Line vs. Off-Line Editing

The reason for the existence of both on-line and off-line editing is a matter of economics. Digital video effects, $75,000 tape machines, sophisticated switcher effects, and character generators drive up the cost of the **on-line** edit suite. Yet these same high-priced pieces of equipment often sit idle for much of the time while decisions are made. On-line editing suites can be very expensive to rent, ranging from $200 to more than $1000 per hour, so it makes sense to spend as little time as possible in the on-line suite compiling the project. In the day-to-day post-production environment, much of the time required to edit a 30-second spot or a 2-hour television drama is spent making rather subtle decisions regarding when, where, and how long to make a shot. Because the decision-making phase of editing usually does not require the sophisticated hardware that makes the suite so expensive to operate and rent, it makes sense to make the decisions before renting time in the suite used to assemble the actual production.

ON-LINE: An edit suite or session in which the resulting product is the finished edited master videotape. On-line edit suites typically feature high-end VTR formats, allow sophisticated effects, and demand a higher rate than off-line suites.

OFF-LINE: An edit suite or session in which the resulting product is an EDL, which in turn is used to expedite the on-line editing of the master videotape. Off-line edit suites are used to make creative decisions and to create a rough cut rather than the finished product.

EDL: Edit decision list. This is a list of numbers signifying in and out points for each of the series of shots making up the program. The EDL may be a written list of numbers, a computer print-out of time code numbers, or a computer floppy disk containing the same list of numbers in a format readable by the on-line editing computer.

The sensible solution to the problem is known as **off-line** editing. Off-line editing is usually done in a simple editing suite, often using a videotape format that is more cost efficient than the format on which the program material originated. For instance, a television commercial shot on film or 1-inch videotape or a documentary shot on Beta-cam could be dubbed to either 3/4-inch or even VHS videotape for off-line editing in a simple cuts-only edit suite. Once the creative decisions have been made, the rough cut and the **edit decision list (EDL)** are used to direct the on-line edit session that uses the original 1-inch or Betacam source footage.

Be prepared to consider the economic advantages and disadvantages of the off-line process. The more sophisticated the off-line suite, the more expensive the hourly rate, but the more complete the EDL, and the more detailed and accurate the EDL, the fewer hours spent in on-line editing where the costs add up even more quickly.

The EDL that results from an off-line session may be as simple as a hand-written list of time code numbers taken from the visual time code displayed on the source tapes, or it may be a computer-generated EDL created by the computerized off-line editor and printed out by the editor's printer. Sometimes, the result is a computer floppy disk of the EDL, using time code numbers, that can be loaded directly into the on-line editor. Many high-end computer editors create CMX-compatible EDLs and floppy disks, but depending on the system you choose, you'll need to verify the format of the EDL and its compatibility with the on-line edit system you'll be using.

Whether you leave the off-line edit session with a **paper edit** or a floppy disk containing your EDL, the next step is to perform the actual program assembly in the on-line edit suite. At this stage, the editor and producer can concentrate on the technical quality of the edit rather than on making creative decisions. Remember that whether a suite or session is on-line or off-line really has little to do with the format of the VTRs used but rather whether the finished product is the finished edited master videotape. If the result is the finished tape ready to air, it is an on-line session and suite. If the result is a rough-cut, or intermediate, master with its EDL, the session and suite is considered to be off-line.

PAPER EDIT: This phrase is used to describe the off-line editing process in which the producer keeps a hand-written log of the edits and their in and out points. This can be performed using a simple cuts-only editing system and window dubs of the time-coded source footage.

Time Code

For most post-production activity, an issue of primary importance is being able to accurately access a particular frame or section of the recorded videotape. For years, the way to do this was to use electronic counters that read control track pulses recorded on the videotape. Control track, as you may remember, is the electronic pulse recorded along with the video and audio signals on the videotape. Because it is simply a pulse, it does not have the capacity to store or impart information. Control track counters simply count pulses as the tape moves across the heads. At any point on the videotape, the control track counter can be reset to begin reading again from 0:00:00:00.

Time code, on the other hand, is an audio signal that stores and supplies information. A specific numerical value corresponds to each frame of video. Because time code provides an assigned value for a specific frame of video, a time code reader will read a discrete number for each frame.

One advantage of time code, then, is the ability to note a particular frame or segment of a videotape and recall that time code value on the same or a different VTR at a later time. For videotape editing, time code allows repeatability (i.e., an edit list can be created, stored, and recalled to perform the exact same edits at a later time). Another advantage of time code over control track is accuracy. Time code editing is accurate to the frame, and some systems even provide single-field accuracy, while the best control track systems typically provide accuracy of ±two frames. It is interesting to note that a time code address for an edit actually specifies the point between field 1 and field 2 of a particular frame of video. A match-frame edit shows up in the EDL with the same frame address for the in point as the previous out point, with the edit taking place in the middle of that particular frame of video. There are several types of time code, but the recognized standard for broadcast applications in the United States and in Europe is SMPTE/EBU time code.

Drop Frame vs. Non–Drop Frame Time Code

As mentioned in earlier chapters, the frame rate for NTSC video is actually 29.97 fps rather than the 30 fps figure more commonly used for general description of the NTSC system. As you might guess, this causes some complications for a time code system based on 30 fps. Because each and every frame of video receives a time-coded number, the approximate value of 30 fps of time code must somehow be resolved with the actual value of 29.97 fps. The means devised to correct this discrepancy of 3.6 seconds per hour is known as drop frame time code. For the time code to match actual clock time, two frames of time code are dropped each minute, except for every tenth minute. In practice, then, the frames of video would be numbered as follows:

> 1:00:59:25 or 1 hour, zero minutes, fifty-nine seconds, and twenty-five frames
> 1:00:59:26 or 1 hour, zero minutes, fifty-nine seconds, and twenty-six frames
> 1:00:59:27 or 1 hour, zero minutes, fifty-nine seconds, and twenty-seven frames
> 1:00:59:28 or 1 hour, zero minutes, fifty-nine seconds, and twenty-eight frames
> 1:00:59:29 or 1 hour, zero minutes, fifty-nine seconds, and twenty-nine frames
> —Drop two frames (frames 1:01:00:00 and 1:01:00:01)
> 1:01:00:02 or 1 hour, one minute, zero seconds, and two frames
> 1:01:00:03 or 1 hour, one minute, zero seconds, and three frames

The difference between time code and clock time, if these frames were not dropped, amounts to an additional 3.6 seconds per hour. For this reason, drop frame time code is frequently used when editing program length productions for broadcast. This allows the editor to be constantly aware of the exact duration of the show in progress. On the other hand, there may well be times when non–drop frame (NDF) time code may be a more logical choice. A word of caution: Editing of material for interactive videodiscs should always be done with the master tape striped with NDF time code. The reason for this is that the computer software is programmed to access each individual frame and could be confused if it encountered a missing frame number.

Longitudinal vs. Vertical Interval Time Code (VITC)

Time code is an audio signal, and like any other audio signal, it can be recorded on videotape or audio tape in several different ways. The most common way is to record the signal with a stationary head along the length of the tape. This is known as **longitudinal time code**. The other option is to write the information as part of the video signal, or to be more precise, within the vertical blanking interval of the video signal. This is known as **vertical interval time code**, or **VITC** (vit•see). Each word of longitudinal time code is made up of 80 bits, and 90 bits are used for the VITC signal.

LONGITUDINAL TIME CODE: Time code that is recorded on the videotape on a linear audio track rather than in the vertical blanking interval of the video signal. Longitudinal time code is an audio signal that uses 80 bits to assign a numerical value expressed in hours, minutes, seconds, and frames to each and every frame of video.

VITC: Vertical interval time code. Time code that is recorded within the vertical blanking interval of the video signal. VITC uses a 90-bit word to assign a value to each frame of video.

Longitudinal time code and VITC each have their advantages and disadvantages. Longitudinal time code can be added to a videotape after the program audio and video signals have already been recorded. Another advantage of longitudinal time code is that it is easier to read accurately at high tape speed (i.e., when the tape is being shuttled). VITC, because it is part of the video signal, cannot be added after the video is recorded unless you choose to dub down a generation. However, because it is written and read by rotating heads, VITC can be read while the videotape is in slow motion or even when still. Longitudinal time code, because it is susceptible to deterioration from generation loss, must be regenerated by a jam-sync time code generator when it is dubbed from one tape to another. However, VITC usually dubs without any problems. Some editing systems can use both longitudinal and VITC, taking advantage of the benefits provided by each.

Window Dubs

Some time code generators and readers can generate video characters from the time code signal. These characters can be superimposed over the video signal to allow visual, on-screen identification of the time code signal. These dubs with visual time code are known as **window dubs, in-vision time code dubs,** or **burned-in time code.** Source footage shot on 1-inch or Betacam SP, for instance, can be dubbed to VHS with time code "in-vision" for the producer or director to view at home. Window dubs are commonly used for viewing or off-line editing for the purpose of compiling a paper edit list. Because the visual time code is now part of the visual image, the window dubs are useful *only* for off-line editing. For window dubs to be effective, it is important that the visual time code match frame-for-frame the time code on the original source footage tapes.

WINDOW DUBS (BURNED-IN TIME CODE or IN-VISION TIME CODE DUBS): A dub of a tape showing the time code as visible characters superimposed over the video picture. The visible time code must match the time code recorded on the original source footage for the window dub to be of any use.

Linear vs. Random-Access Editing

Videotape is a linear medium. To get from point A to point B on a piece of videotape, you must shuttle through the tape between the two points. Even with high-speed shuttle on today's VTRs, the time it takes to get from one point on the tape to another is time lost in the editing process. Not only that, but to make a change in the middle of an edited videotape requires that everything from that change to the end be re-edited, unless of course the new material is exactly the same length as the replaced material. Editors have for decades been wishing for a system that would allow them to work as fast as they can think—one that would allow updates and changes without the hassle of re-recording. To be able to select a sequence of shots and then have the machinery display the results immediately is every editor's dream. Videotape, at least in the way that it is normally used in **linear** editing, demands that the editor wait on the machines.

However, there is an alternative, and it's known as **random-access**, or nonlinear, editing. Consider the difference between an audio cassette and a phonograph album. To get from the first song to the third song on side A requires different techniques and takes a different amount of time. The audio cassette must be fast forwarded, but the record album allows faster access by picking up the stylus and repositioning it closer to the center.

LINEAR: Having to do with recording formats or editing processes in which the material is accessed by shuttling forward or reverse. Magnetic tape is a linear medium in that it requires that the operator search in a sequential fashion.

RANDOM ACCESS: Having to do with formats or processes in which the material is accessible randomly and in the same amount of time. Random access of audio or video material is achieved by any of the disk-based formats, e.g., phonograph, CD, laser disc, or computer hard disk.

Random-access editing means that instead of recording the selected takes onto videotape, the nonlinear editing system simply builds an edit list in memory that allows all of the selected shots to be recalled in real time for a preview. Because the edited material does not exist in its assembled form on videotape, changing and updating it is as simple as adjusting in and out points in the EDL. In a nutshell, that's the difference between linear and random access.

For videotape editing to become a random-access procedure requires one of two approaches, both of which are currently in use. The first is to use a bank of multiple VCRs, each VCR loaded with a dub of the source footage. In reality, this system provides pseudo-random access because it still deals with source footage on a linear storage medium, i.e., videotape. The computer editor keeps track of where each VCR's tape is positioned so that when the editor asks to see a particular shot, the computer selects the VCR with the tape closest to that position. This allows fast cuing and permits multiple shots to be replayed before recording to the edit master tape. As one machine is playing back, several other VCRs are cuing up in preparation to play back. As the VCRs play, the computer automatically switches between the source machines in the proper sequence. The Ediflex by Cinedco and the Montage Picture Processor are just two of the systems that currently use this approach to achieve random access.

The other technique, and the one with greater promise because it provides true random access to both the source footage and the edit master, is to compress and transfer the video material to either optical disk or computer hard disk. Because both optical disks and computer hard disks provide random access, editing becomes a much more responsive process. The CMX 6000, Editing Machines Corporation Emc2, Montage III Digital Picture Processor, and the Avid Media Composer are just a few examples of disk-based editing systems. The Avid systems use the Macintosh® personal computer as an interface, using the mouse and familiar icons to perform editing functions.

One important note to add to all of this: Random-access editing is currently available only for the purpose of creating an EDL and not a finished edited master. Currently, no on-line random-access editing systems are available. However, with advancements in video compression, the quality of compressed video is expected to increase, making broadcast-quality output from random-access systems a reality.

It is interesting to note that Lucas Films' disk-based Editdroid system failed to achieve initial success and was removed from the market for some time. Recently, however, it has begun to make a comeback and is once again on the market. Likewise, the Montage company declared bankruptcy in 1986 but has come back stronger than ever. It would appear that the transition from linear to nonlinear editing, although ripe with promise, may for the next several years be a rocky road for manufacturers and users alike.

Videotape Editing System Design

Videotape editing, at its most basic level, can be accomplished with a source of video and a recorder with flying erase heads. Flying erase heads make possible a clean record transition between two shots. A camcorder with flying erase heads can be started and paused, and each new shot will be cleanly edited onto the end of the previous shot. The finished product is edited in camera.

The next step up is to use two VCRs, one as the source machine and the other as the editing VCR. The source machine must be capable of playing back the previously recorded tape, and the record machine must be able to perform edits. Once again, the flying erase head allows clean edits between shots. A recording VCR without flying erase heads will still allow you to record a series of shots in order; however, the transition between shots will not be clean, i.e., there will be breakup, instability, or video snow at the edit points. If the editing VCR does not provide machine control of the source machine, it will be necessary to cue the source and record VCRs manually and to roll them together to make the edit. This can be a frustrating process, especially when you are trying to make precise edits within either the video or audio tracks.

If the record VCR is an editing VCR, it will allow you to select either assemble or insert editing. If you select insert editing, you have the choice of editing any or all of the following tracks: the video track, audio track 1, audio track 2, and audio tracks 3 and 4 (if applicable). Some VCRs even allow editing of the time code and control tracks. Basic editing systems can be assembled using an editing VCR that controls the source machine.

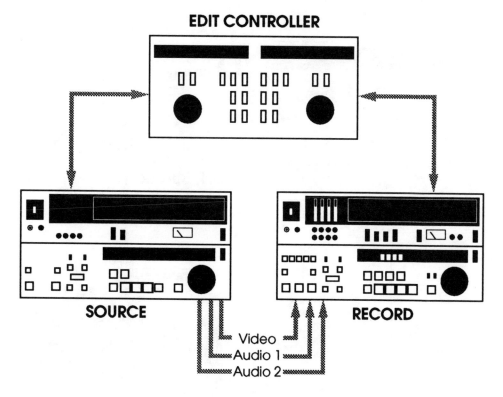

Basic Editing Suite Configuration

This machine-to-machine control is enabled by using a multipin cable connecting the two machines. In this configuration, the edit VCR is used to set the in and out points for both machines and allows automatic preview, edit, and review functions, and it controls the tape transport functions for both VCRs. This configuration is used for down and dirty news editing on location or in a mobile truck.

The next step up from this simple two-machine system is a basic cuts-only editing system. The components include a source VCR, an editing VCR, an edit controller, and of course the appropriate audio and video monitors. Otherwise known as *butt editing*, this system permits cuts only—no dissolves, wipes, or other video transitions. Most of these systems allow precise edits if the VCRs use time code; otherwise, they provide an accuracy of ±two frames using control track.

Although cuts-only editing may suffice for the simplest post-production tasks, frequently more capability is required. This is achieved by adding another source machine to create an A/B roll editing system. The first source machine is the A machine, and the second is the B machine. Effects such as dissolves, wipes, and other switcher transitions can now be included in the editing process. However, the second source machine is only one of the items necessary to convert a cuts-only system to A/B roll. The source machines will have to be time base corrected to permit accurate timing and glitch-free transitions at the video switcher. Also, the edit controller must be capable of controlling at least three

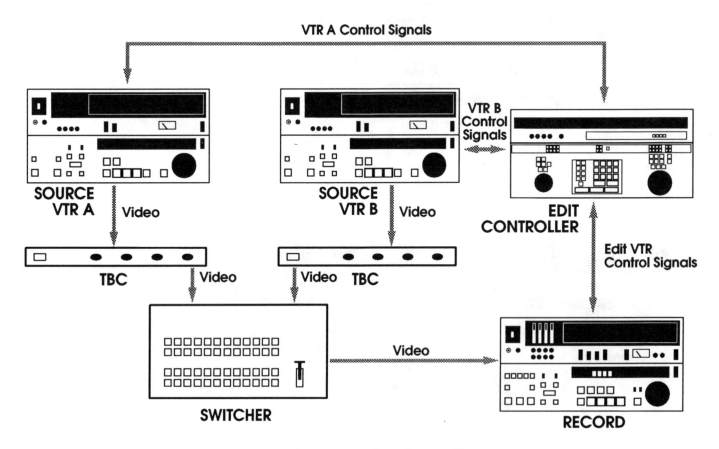

A/B Roll Editing Configuration

machines for effective editing. A video switcher will need to be added, although some A/B roll editing systems are being introduced that combine switcher and TBC in the edit controller. This may work just fine 90% of the time, but this approach permits less flexibility than a system built from components. Finally, time code is no longer simply a luxury; with A/B roll editing, time code becomes a functional necessity. Note that the D-2 format digital videotape recorder is the first to employ write-after-read heads. This feature allows A/B effects to be performed using only one source machine with the D-2 recorder.

From this point on, the expansion of the edit suite is simply a matter of adding additional source VTRs, digital video devices such as the **Ampex Digital Optics (ADO)** unit or Quantel's Mirage system, and digital recording devices such as the Abekas A-62 digital disk recorder. **General purpose interface (GPI)** triggers can be used to remotely start audio tape decks and CD players for additional audio sources in the edit suite. Almost all on-line editing suites have some type of character generator available, and a few have paint systems as well.

ADO: Ampex Digital Optics. This is probably the most popular digital video effects device in use in post-production editing suites. *ADO* is sometimes used synonymously with *digital video effects.*

GPI: General purpose interface. This is a computer editing protocol used to trigger other devices in the edit suite, such as the ADO, the switcher, or an ATR. The GPI is a parallel interface that can be set to trigger a device at a particular moment during an edit. The exact moment is usually defined by time code number.

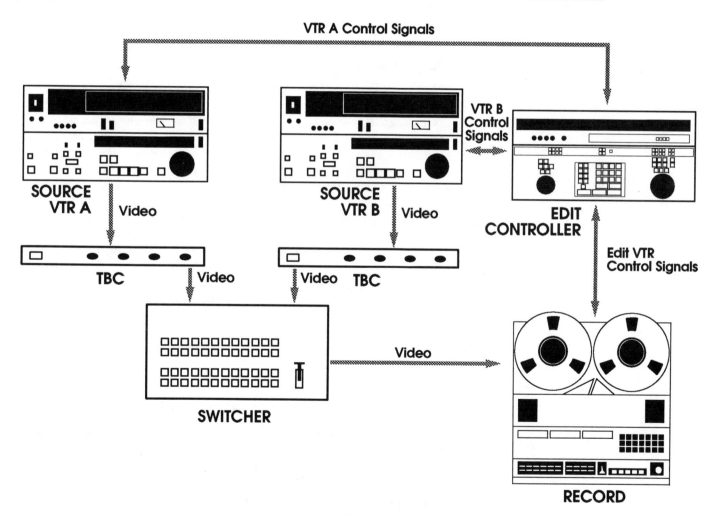

Interformat Editing Configuration

Interformat Editing

It is quite common these days for an editing suite to incorporate various formats of tape machines. The reason for this is that producers are choosing to record their programs on several viable formats. To serve the widest possible client base, edit suites are trying to accommodate various formats. More often than not, the record machine will be a standard broadcast format; the most common format today is 1-inch type C. It is important that the edit master be recorded and delivered on the format stipulated by the client. A corporate video may be mastered onto 3/4-inch tape for playback on their house system, while a television spot is typically mastered onto 1-inch, as are most other products for broadcast. However, the suite may have source machines capable of handling various formats, including D-2, 1-inch, Betacam SP, M-II, 3/4-inch, and even the S-VHS and Hi-8 "prosumer" formats. Another convenience of the interformat suite is that tapes can be dubbed from one format to another, assuming that each of the videotape machines is capable of recording as well as playing. A VHS or 3/4-inch dub is sometimes needed to show to someone whose approval is necessary for continuation of the project.

Posting Film on Videotape

Strange as it may seem, one thing that the tediously slow process of developing a high-definition television system has brought about is an increased vitality in the film-for-television business. Television program producers and directors, who are unsure of where video technology is headed but know that it is certainly headed somewhere, are playing it safe by continuing to shoot on 35mm film stock. The theory behind this decision is that no matter what high-definition television system is finally selected for broadcast and distribution, their 35mm master can be transferred to the new video standard while maintaining the highest quality possible. Although much of the film that is shot is also posted on film, quite often the film is transferred to videotape before editing. This film-to-videotape transfer, whether performed on the entire lot of original film footage, only on selected shots, or on the finished film master, is typically performed using a piece of equipment known as a *flying spot scanner*. Two of the leading names in flying spot transfer technology are Rank Cintel and Bosch. Today's flying spot scanners offer a vast improvement over earlier film-to-tape transfer processes, such as film chains or telecines. Whether the original film footage is transferred to D-1, D-2, 1-inch, or one of the professional 1/2-inch formats, the transfer process must be carefully monitored to preserve the high quality often associated with 35mm film.

Component vs. Composite Suites

Most videotape editing takes place in the composite video realm. Regardless of whether the VTRs are composite video recorders (1-inch, D-2, or 3/4-inch), component (D-1, Betacam SP, or M-II), some other variation of component VTRs (S-VHS, Hi-8, or ED-Beta), or a combination of VTRs from each category, most suites process and route the video signal in its composite format. One advantage of the composite suite is that the routing of the video signal requires only one wire instead of three for the component video signal. However, component edit suites do exist and for good reason. The advantages gained by keeping the video signal in its component form while recording are best maintained by keeping the signal component throughout the entire editing process. Component switchers, character generators, TBCs, distribution amplifiers, and digital effects devices make this possible. The signal must be encoded sometime before broadcast, but in most circumstances the rule of thumb is the later, the better. In the ideal system, the entire station or plant would operate in the component realm with one high-quality encoder positioned in line just before the signal goes to the transmitter or to another form of distribution.

Technical Considerations

Color Framing

As already mentioned, the NTSC system uses two fields to make up each frame of video. When working with a full-bandwidth color video signal such as those recorded on 1-inch videotape, this same system requires four fields to make up a **color frame** due to the alternating phase of the subcarrier-to-sync relationship. Fields 1 and 2 are defined as color frame A, and fields 3 and 4 are defined as color frame B. This alternating A-B-A-B sequence must be maintained during the editing process. If an edit is performed without regard to maintaining the integrity of the color framing, a horizontal shift of approximately 1/227 of the picture's width or another inconsistency can occur at the edit point. A horizontal shift as small as this may seem trivial, but it is noticeable when doing match-frame edits. For this reason, most computerized editing systems have automatic color framing. Some systems use a color field identification pulse (CFID) that is added to the control track pulse. Others use the time code signal to identify color frame A (even time code number) from color frame B (odd time code number). It is standard procedure in either case to set the color burst phasing for a particular reel of tape when it is loaded. This setting should remain unchanged throughout the entire reel. When an in point is selected that would violate the color framing sequence, the edit controller automatically adds or subtracts one frame from the in points of both the source and the record VTRs, thus maintaining the proper color framing for the edit.

Generation Loss

Working in the analog world, video and audio post-production is constantly fighting the effects of generation loss—the inherent signal quality loss that results from repeated copying from one tape to another. Generation loss, which is most noticeable as increased noise and velocity error, is inevitable when dealing with analog signals. However, it does not affect all formats equally. One-inch type C video recorders typically deliver up to seven or eight generations of broadcast-quality images before the signal becomes noticeably degraded. The professional component 1/2-inch formats are said to be able to handle five to six generations, while 3/4-inch video is usually held to three to four generations by professional broadcasters. Some VTR manufacturers have developed high-end time base correctors that have a look-ahead feature that shows accumulated error of the video signal. This feature allows the tape operator to set levels and parameters much more precisely; the result is more generations before the signal becomes degraded. The Ampex Zeus video processor, in conjunction with the Ampex VPR-3 1-inch VTR, has been demonstrated to provide more than 20 generations with results almost identical to the normal seven to eight generation limit for this format.

Analog audio recorders experience similar problems with increased noise with each generation. With professional audio tape recorders, the noise typically increases 3 dB with each dub. Increased tape speed and noise reduction processes are used to contend with the problem but with only limited success.

Currently, the best solution to the generation problem is digital recording and processing. Theoretically speaking, digital recording and dubbing avoids all of the normal generational loss associated with analog recording. In practice, digital recording permits up to a hundred generations or more of audio or video recording with acceptable results. The best results are obtained by using hard-disk recording technology for multigeneration projects; however, digital audio tape and videotape recorders provide similar advantages on a more limited scale.

Audio Post-Production

Audio post-production ranges widely in terms of the technology employed and the sophistication of the process. Simple audio editing can be accomplished using a 25¢ razor blade, a grease pencil, and a splicing block, while high-end audio post-production for television or film can entail months of work with digital multitrack audio tape recorders and random-access disk-based digital editing systems.

Audio post-production for video production is known as *audio sweetening*. Audio sweetening, like so many other post-production processes, is heavily dependent on time code. In this case, time code is what makes it possible to transfer the audio track from the edited master videotape to a multitrack ATR, where the sweetening process is performed, and then back to the edited master videotape—all this while retaining a synchronous relationship with the visual images. Time code on both the VTR and the ATR makes it possible to sync up both machines during the lay-over and lay-back process. Once the audio has been transferred to the audio tape recorder, multiple tracks become available for music, sound effects, and multichannel audio program material. If musical instrument digital interface (MIDI) technology is being used, time code readers can be locked to MIDI signals from the host computer, thus automating the lay-back step of the process.

Self-Study

Questions

1. This type of editing requires that you begin with an edit master videotape that has been prerecorded with control track (also commonly known as a blacked tape):
   a. control track
   b. assemble
   c. insert

2. The Avid Media Composer and the Ediflex are two editing systems that allow this type of access to source footage during the editing process:
   a. linear
   b. longitudinal
   c. random access

3. A simple A/B roll editing system requires a minimum of this number of VTRs:
   a. two
   b. three
   c. five

4. Drop frame time code is based on the fact that one second of video is actually made up of this number of frames of video:
   a. 30
   b. 60
   c. 29.97

5. This part of the video post-production process is concerned with the generation of an EDL and not the finished edited master videotape:
   a. off-line
   b. on-line
   c. logging

6. The several seconds of time required for the videotape machines to roll up to the edit in point, allowing the machines to synchronize, is known as:
   a. jam sync
   b. preroll
   c. post-roll

7. This type of time code is written as part of the video signal and, as such, can be read by the edit controller even while the tape is paused or moving at very slow speed:
   a. VITC
   b. longitudinal time code
   c. drop frame time code

8. Of the professional analog videotape recording formats, this one provides the best performance and the least susceptibility to the effects of generation loss:
a. Betacam
b. BVU-SP
c. 1-inch

9. This device is necessary to correct the instability of a video signal that originates from a videotape machine; it is also necessary for each playback machine in an A/B roll editing system:
a. time code reader
b. time base corrector
c. sync generator

10. This is used to allow the edit controller to trigger an external device, e.g., an ATR or digital effects box, during the editing process:
a. DVE
b. TBC
c. GPI

Answers

1. a. No. Control track editing may be performed on either a blank tape or one that has been blacked.
b. No. One of the few advantages of assemble editing is that it does not require that you have a blacked tape before beginning to edit.
c. Yes. Insert editing allows you to record on either the video track or any of the audio tracks, but it does not affect the control track.

2. a. No. Linear access to source footage is provided by videotape, which unfortunately must be viewed, played, or shuttled in a linear fashion.
b. No. Longitudinal, which is very similar in meaning to linear, is the type of access provided by videotape-based editing systems.
c. Yes. Although the two systems achieve random access in two different ways, each system provides nearly instantaneous access to any section of source footage.

3. a. No. Two machines, one playback and one record, are the minimum number required for a simple cuts-only editing system.
b. Yes. Two source machines and one record machine are required for A/B roll editing.
c. No. Although an A/B roll system may use many videotape machines, only three are required to record A/B roll effects, e.g., dissolves and wipes.

4. a. No. Although 30 is used as an approximate figure, the actual figure is a little less.
b. No. However, there are approximately 60 fields per second.
c. Yes. This slight discrepancy with clock time is corrected by using drop frame time code.

5. a. Yes. Off-line editing is used to generate the EDL, which is then taken to the on-line session that results in the finished edited master videotape.
b. No. The on-line editing process is concerned with the actual assemblage of the program and the creation of the finished edited master videotape.
c. No. Logging usually refers to the viewing of source footage and the recording of time code values for edit in and out points.

6. a. No. Jam sync is used to describe the process of regenerating time code.
b. Yes. For most editing systems, a preroll of 5 seconds is standard, although other values may be used.
c. No. Post-roll is in fact the amount of time the videotape machines continue to roll after the edit out point has been reached.

7. a. Yes. Vertical interval time code is actually written in the vertical blanking interval of the video signal and is read by the rotating video head drum.
b. No. Longitudinal, as its name implies, is written on a linear track and is read by a stationary head. For this reason, the longitudinal time code signal cannot be read unless the tape is moving past the head.
c. No. Drop frame time code can be recorded as either longitudinal time code or VITC.

8. a. No. Although the manufacturers of Betacam SP products would like to believe that Betacam's performance matches 1-inch, it doesn't stand up as well to the same number of video generations.
b. No. Of the formats listed, BVU-SP, which uses 3/4-inch tape, has the worst performance over multiple generations.
c. Yes. One-inch, especially with one of the advanced time base correctors available, still has the best generational performance of any analog VTR. Digital VTRs are another story.

9. a. No. The time code reader converts the time code signal to alphanumeric characters.
b. Yes. The TBC is a digital device that corrects time base instability and makes possible the mixing of videotape signals by a switcher.
c. No. Although a sync generator generates stable sync pulses, it does nothing to correct instable video from a videotape machine.

10. a. No. DVE is a trademarked name for a digital effects system manufactured by the NEC Corporation.
b. No. The time base corrector does not control external devices; it merely processes video signals from a videotape machine.
c. Yes. The general purpose interface is used by edit controllers to trigger (i.e., start) external devices; however, it does not provide the level of control needed for the VTRs involved in the editing process.

Projects

Project 1

Take a field trip to a video post-production facility. The facility may be as simple as a videotape editing bay in a local TV station's newsroom, or it may be a sophisticated post-production house with the latest tape formats and digital video effects. In either case, take this opportunity to observe the configuration of tape machines, monitors, and control devices. Ask for a brief demonstration of the various types of editing performed there. Compare the off-line editing bay with the on-line suite.

Purpose

To observe the design and configuration of an editing facility.

Advice, Cautions, and Background

1. Remember that video post-production is an extremely capital-intensive business. The amount of money tied up in an on-line suite can be astronomical. Be prepared for a very brief tour of the on-line suite if you get the opportunity to see it at all.

2. Editors sometimes are a very interesting mix of personalities and abilities. Editing requires both an aesthetic and creative approach to the job and demands technical expertise as well. That's not all; an editor must also be very good at working with people. For these reasons, an editor can be an extremely valuable employee, particularly in the competitive video post-production business.

Chapter 8
SIGNAL MONITORING

MONITOR: Audio: A transducer for converting electrical energy from an amplifier into sound energy. **Video:** A transducer for converting electrical energy into visible light.

The monitoring of a signal, whether audio or video, is at once both a critical and a subjective process. Although every part of the audio and video signal process is important, it is the **monitor**, either an audio loudspeaker or a video CRT, that ultimately determines what the audience hears and sees. Audio speakers and video monitors may attempt absolute transparency, but never actually achieve it. That is to say, the monitoring devices themselves introduce information not present in the audio or video signal alone. Although some of this is unavoidable and expected, the bottom line is that a certain degree of subjectivity is associated with the monitoring process. For broadcasters, perhaps the greatest variable of all is the inability to control the monitoring equipment and the environment of the end user. What the musician hears in the studio control room and what the television director sees on the control room monitor may be very different from what the consumer experiences at home. This is the nature of the world in which the audio and video professional must operate. Most would agree that the highest degree of control is available to the film director. Due to the limited number of theaters and the partially controlled environment provided therein, the theatrical film experience has the greatest potential to closely approximate what the film director has envisioned. Be that as it may, you should begin by considering the technology at work in audio monitoring equipment, more commonly referred to as *audio loudspeakers*.

Audio Monitors

Audio monitors convert the amplifier's electrical energy into mechanical or acoustic energy. The reverse of this process has already taken place at the origination of the sound source (the microphone converted acoustic energy into electrical energy). However, the higher power levels and volume required of loudspeakers make the task more complicated and increase the margin for error. The design of the speaker elements, crossover network, and cabinet all affect the performance of the loudspeaker system.

Loudspeaker cabinet design is both an art and a science. To achieve optimum accuracy and maximum efficiency, designers have come up with a plethora of shapes, sizes, and schemes for their speaker cabinets. Three of the more popular are the acoustic suspension, bass-reflex, and satellite/subwoofer systems. The acoustic suspension design uses a sealed cabinet. With this design, the bass response comes primarily from the forward motion of the speaker cone. Generally speaking, larger enclosures are needed to obtain the lowest frequencies when using an acoustic suspension design. In contrast, the bass-reflex cabinet has ports, or tuned holes, in the cabinet to enhance the bass response. This provides much greater efficiency. That is to say that the bass-reflex speaker provides greater volume than an acoustic suspension speaker using the same amount of power. The trade-off, however, is a matter of reduced accuracy and diminished control of the signal. The third type of design, the satellite/subwoofer system, uses separate enclosures for the middle to high frequencies and the bass frequencies. The subwoofer enclosure produces low, nondirectional frequencies, and the satellite speakers are carefully located to provide optimum response and imaging. The flexibility in speaker placement makes this last option especially attractive both to the audio designer working with limited or unusual space and the consumer audiophile.

Once the cabinet is designed, the selection and placement of speaker elements is critical. The most common type of loudspeaker element is the moving coil design. Moving coil speakers function just like a dynamic microphone but in reverse. However, unlike a microphone diaphragm, a single speaker element cannot efficiently reproduce both extremely low frequencies and extremely high frequencies. For this reason, most speaker cabinets combine two or more speaker elements. These are commonly known as *two-way* or *three-way* speaker systems. The primary types of speaker elements are **woofers**, **horns**, and **tweeters**. Woofers dispense the lower frequencies (or bass notes), horns handle the mid-range, and tweeters reproduce the highs.

For the woofers, horns, and tweeters to work together with maximum efficiency, crossover networks are used to separate the audio signal into these three frequency ranges. Then, these three separate signals are used to drive only those speaker elements designed to reproduce those frequencies. These crossover networks can be either active or passive devices. The difference between the two is that active crossovers split the signal *before* amplification, thus allowing the amplifiers to be matched to the frequencies being processed. This approach is considered to be more efficient and results in improved performance. In the passive crossover system, the signal is amplified first and then split into two or more frequency bands that are routed to their respective speaker elements. Some powered monitors have the amplifiers, crossover network, and speaker elements all in one cabinet. The Electro-Voice Sentry 100 EL is an example of such an arrangement.

In many broadcasting environments, such as studio control rooms and editing suites, the operator has a choice of monitoring speakers. The operator can select very high quality reference monitors with extremely flat frequency response, thus enabling him or her to hear exactly what the audio signal is doing at all times. The operator may also select a smaller, lower quality set of audio monitors to more closely approximate the sound that the end user (radio or TV listener) is going to experience. The rationale behind this is that it is considered to be counterproductive to monitor audio only on $5000 studio reference monitors when the listener at home or in the car will likely be listening to the same audio on inexpensive 4-inch speakers. No audio producer or engineer willingly chooses to mix audio while listening to monitor loudspeakers that mask or color the true audio. However, it does make sense to make mixing decisions that take into account what the majority of listeners are going to experience.

Speaker Phase

When wiring loudspeakers, i.e., connecting the amplified signal to the input terminals of the speakers, it is important that the loudspeakers be wired in phase. The amplified signal has a positive and a negative lead, and these must be connected to the positive and negative leads of the loudspeaker. If one speaker is connected properly and the other is wired with the leads reversed, the speakers are wired out of phase. The result is that one speaker's elements will generate a positive sound wave while the other speaker generates a negative wave. The effect is a slight, although potentially serious, signal loss for the listener. Almost all speaker wire is coded, and it is a fairly simple matter to be sure that the positive and negative leads are connected properly at both the amplifier and the loudspeaker.

Video Monitors

The television picture monitor is where the end result of the video production process is seen. It is here that the images that were once captured by the camera, recorded on the VTR, processed through a TBC, and perhaps broadcast by microwave, satellite, or standard RF transmission are now viewed by the audience. What they see may be an accurate reproduction of the original scene, or it may be quite different. You must assume, however, that the average television receiver is reasonably adjusted and capable of displaying at least an approximation of what is sent to it. But how do you ensure that what you see all along the production process is an accurate display of what you have on tape or in the viewfinder? How much faith should you put into the one piece of equipment that many consider to be the weakest and most subjective link in the video production process? Well, what it boils down to is using the best possible reference monitor your budget can afford, making sure that this monitor is carefully maintained, relying also on the waveform monitor and vectorscope, and finally, having a little bit of faith.

Before going too far, consider the difference between a television receiver and a television monitor. Essentially, the difference is that a receiver contains a tuner capable of receiving television broadcast signals (RF signals) and down-converting them to audio and video signals for display. A monitor, on the other hand, does not contain the electronics necessary to receive broadcast signals but has the proper connectors to allow video and sometimes audio signals to be connected to the monitor for display. A reference, or broadcast-quality, monitor is a precision instrument capable of displaying an accurate, high-resolution picture. It allows television technicians to make reasonably accurate judgments based on viewing the display. On the other end of the spectrum, video monitors that are used primarily for display purposes need not be so precise. In fact, it is often better that they mask or hide imperfections in the video signal being displayed.

Video monitors and receivers must process complex electronic signals, including picture information and synchronization signals. The NTSC video sync signals direct the monitor in converting the electronic signal into an orderly representation of the original camera scene. Horizontal and vertical sync pulses convert a continuous stream of electronic pulses into individual lines of picture information. These lines in turn make up fields, which make up video frames, which are displayed at a rate of 30 times each second to produce full-motion color video images.

Another frequent use for video monitors is as computer displays. It is important to understand that the sync rate for computer video displays is almost always different from NTSC video sync rates. The video display connected to a PC probably cannot display NTSC video, just as most TV receivers cannot be used in place of a computer video terminal. Several manufacturers have recently introduced monitors that will adjust automatically to a variety of sync rates, making them capable of displaying NTSC video or any number of computer video signals. One manufacturer's automatic scanning capability allows operation at any horizontal scanning frequency between 15.5 and 38 kHz and at any vertical scanning frequency between 50 and 100 Hz. Another complicating factor is that many computer displays use sequential scanning, and the NTSC video signal uses interlace scanning. When dealing with video, remember that not all video is NTSC video.

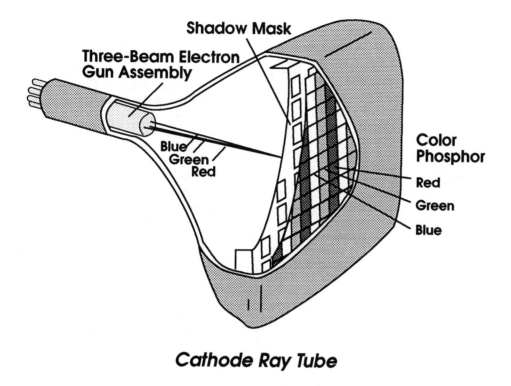

Three-Beam Electron Gun Assembly

Shadow Mask

Blue
Green
Red

Color Phosphor

Red

Green

Blue

Cathode Ray Tube

How a Cathode Ray Tube (CRT) Works

The video monitor is first and foremost a large vacuum tube, or more precisely, a cathode ray tube (CRT). Within a monitor's CRT, electron guns shoot electronic beams—one beam for each of the primary colors (red, green, and blue). In reality, the beam is a stream of electrons that have been boiled off the cathode and that are accelerated toward the phosphor-coated screen. These electrons strike the back of the picture tube, which is coated with red, green, and blue phosphors. When the beams strike the phosphors, the phosphors fluoresce, or glow, producing visible color on the screen. The strength of the beam as it strikes each phosphor dot determines the brightness or luminance of that particular pixel. The red beam strikes only the red phosphors, the blue beam strikes the blue phosphors, and so on. In one of the many miracles of television, the result of this jumble of glowing red, green, and blue dots is a rich and colorful image for the viewer.

Most monitors use some sort of **shadow mask** to prevent beams from striking phosphors of another color. The shadow mask, as its name implies, masks each beam and prevents it from striking phosphors other than those intended. In theory, the green gun's beam strikes only green phosphors, the red gun's beam strikes only red phosphors, and so on. As you might guess, the alignment of guns and masks is a very precise operation.

SHADOW MASK: The metal grill between the guns and the phosphor screen in a CRT. The shadow mask prevents the output from each gun from activating the phosphors of another color.

This technique of using electron guns aimed at color-emitting phosphors is a delicate arrangement. The alignment of the guns and the phosphors, as well as the shadow mask, must be perfect for optimum resolution and color clarity. Several techniques have been used to achieve this goal. The early method, known as *delta-delta,* uses a triangular

DELTA GUN: A CRT design in which the three guns are positioned in a triangular fashion rather than in a line.

IN-LINE GUN: A CRT design in which the three guns are positioned side by side in a line, rather than arranged in a delta configuration.

PULSE DELAY (PULSE CROSS): A feature on some monitors that delays the horizontal and vertical scan so that the H and V blanking intervals are displayed on the screen. Sometimes labeled as *H* and *V delay*. Pulse cross permits the observation of the horizontal and vertical blanking intervals to check pulse widths and system timing. This feature is handy in lieu of a waveform monitor to inspect the blanking intervals.

BLUE-ONLY DISPLAY: A feature that allows the CRT to display the output of only the blue gun. By defeating the red and green signals, accurate settings of hue and saturation can be obtained using a SMPTE color bar test signal. Using the blue-only feature, the operator need not have perfect color vision, simply the ability to discern the differences in brightness values of the bars containing blue information.

UNDERSCAN: A feature on some monitors that permits viewing of the entire picture scanned by the CRT. The overscanned part of the picture is normally hidden by the edges of the display. By using underscan, the operator can check the full area of the picture. This is especially useful when framing camera shots or when using the pulse cross feature.

EXTERNAL SYNC: This feature allows the monitor/scope to receive synchronization pulses from an external source (usually from the house sync generator) instead of taking sync from the incoming video signal. The use of external sync facilitates making timing adjustments.

TALLY: A light, usually red and mounted on either a camera head or monitor, that indicates the "on-air" source.

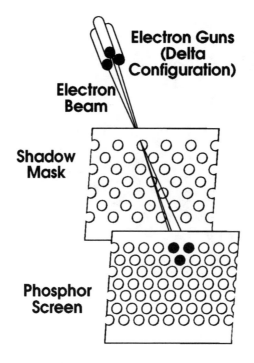

Delta-Delta

arrangement of the guns and the phosphor dots. The **delta gun** monitors were for many years the only CRTs capable of the resolution necessary in a broadcast environment. Unfortunately, delta-delta displays were prone to drift, making it difficult to maintain a sharp picture.

In the 1970s, an improvement in design brought about the in-line/delta monitor. The guns were arranged in a line rather than in triangular fashion. A variation of this design is the in-line/slotted mask CRT. Using an **in-line gun** system with a new phosphor arrangement and a slotted shadow mask between the guns and the back of the tube, this design provides a brighter image. Also, it is available at lower cost, and of course requires much less maintenance than the delta gun monitors. The trade-off, however, is lower resolution because the phosphors are larger, resulting in larger pixels.

Broadcast video monitors typically contain a number of operational features, such as **A-B** switchable inputs, **pulse cross** display, **blue-only display**, **underscan**, reference or **external sync** input, and discrete control of the three guns—red, green, and blue. Some of these features are for the casual user or operator, and some are provided to make the engineer's job a little easier when monitor setup or adjustment is necessary. Another feature of monitors used in broadcast applications is the built-in **tally** indicator. This feature allows the monitor to be wired to crosspoints in the video switcher. When a video signal is selected on the program bus of the switcher, the illuminated tally light indicates that the signal being displayed is the one currently on air. This is especially important in a control room. Whenever you have a bank of monitors displaying various signals from cameras, VTRs, and switcher effects banks, it is important to be able to identify the current on-air signal by noting the monitor with the illuminated red tally light.

Quite frequently, the monitor is the last link in the video signal chain, and as such, it should provide 75-ohm termination. If, on the other hand, you are looping through the monitor to another piece of equipment, the termination switch should be set to Hi-Z. An unterminated signal will appear to be too bright, and a twice-terminated signal will appear dark and dingy.

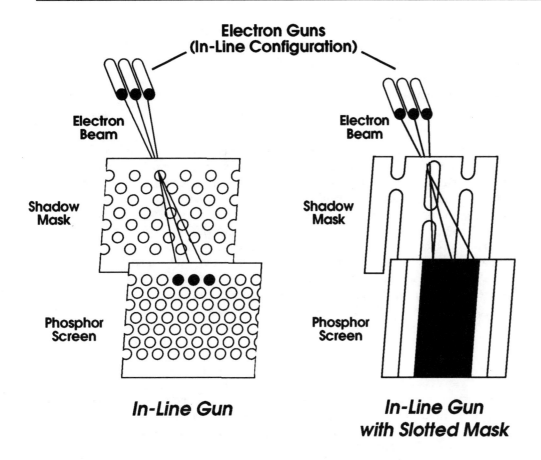

Electron Guns
(In-Line Configuration)

Electron
Beam

Shadow
Mask

Phosphor
Screen

In-Line Gun

Electron
Beam

Shadow
Mask

Phosphor
Screen

In-Line Gun
with Slotted Mask

Red-Green-Blue (RGB), Luminance/Chrominance (Y/C), and Other Component Monitors

Video monitors capable of receiving a composite video signal are the norm, although luminance/chrominance (**Y/C**) and red-green-blue (**RGB**) monitors are available for certain applications. A composite video signal is usually connected to the monitor using a BNC cable, although an eight-pin VTR connector may be used. Y/C monitors have a special four-connector Y/C jack to permit connection to certain VCRs and other Y/C sources. RGB connection is usually supplied by using four BNC connectors, one for each of the separate color signals and sync.

Y/C: Luminance/chrominance. This designation is used for video signals that keep separate the luminance and chrominance information, thus preventing some of the normal NTSC artifacts, e.g., cross-color and cross-luminance. The S-VHS video format is designed to utilize Y/C video signals.

Monitors—You Can't Believe Everything They Say

Anyone who has shot on both film and video knows one of the advantages of shooting on video is the ability to see immediately what the image looks like. Unlike film, which requires processing and printing before an image can be reviewed, video allows for instantaneous playback and monitoring of recorded footage. That is the reason that many film productions now employ video assist. Video assist allows creative people and executives to respond more quickly to questions about performance, composition, and so

155

on. Instead of waiting for the dailies to be screened the following day, they view or review video footage taken from a tap off of the film camera. However, what may first appear to be an advantage of video can quickly prove to be a disadvantage. Although a video monitor is fine for checking composition, framing, and acting, at the same time it is one of the least reliable means of determining the contrast, saturation, and brightness of the image being recorded. As already established, this is the reason for using waveform monitors and vectorscopes alongside video monitors.

Regardless of the questionable value of a picture monitor for critical picture-quality decision, you should make every attempt to maintain your video monitors so as to reproduce images as accurately as possible. Monitors should have a regular schedule for alignment and adjustment. The frequency of scheduled maintenance could range from once a day to once every couple of weeks. After the monitors have been set up they should not be adjusted by the casual user. The reason for this is that if the video technician doesn't like the way the video looks on a particular monitor, the last place to make the adjustment is the monitor itself. That only disguises the problem and does nothing to actually change the video signal. Adding more chroma or contrast to the monitor may make the image more pleasing to you, but it has no effect on the video and the end users will still see the uncorrected video image on their monitors.

Setup Procedures for a Color Video Monitor

How do you go about adjusting a color monitor? Some of the newer studio monitors have computerized setup circuits. Larger facilities may use a color analyzer, a device that is attached to the front of the monitor and that provides an objective reading of its output. However, most people are left to set up their monitors using a color bar signal. If you're lucky, your monitor has a blue-only switch that allows you to view the output of the blue gun only. The procedure is as follows:

1. Allow the monitor to warm up for approximately 30 minutes before proceeding. This will minimize the effects of normal electronic drift. First, make sure that all automatic color or picture circuitry switches are set to the off position. Next, turn the chroma or color control all the way down, so that you now have a grey scale instead of color bars. Now, adjust the brightness control until the black bar appears black, but no farther. On SMPTE bars, the black set or pluge signal provides a convenient reference. Adjust the brightness control so that the whiter-than-black bar (11.5 IRE) is barely visible, and the division between the black bar (7.5 IRE) and the blacker-than-black bar (3.5 IRE) is not visible.

2. Adjust the contrast so that the 100% white bar is quite grey, and then turn it up so that it appears to change from grey to white, but no farther.

3. With the color control set to a medium position, adjust the hue control to achieve colors of the proper shade. If your monitor has a blue-only switch, select the blue gun only; otherwise, use a blue filter to view the monitor. Adjust the hue so that the bars containing blue (white, cyan, magenta, and blue) all appear to have the same intensity. If you're using SMPTE bars, the cyan and magenta bars will match the smaller bars just below them.

156

4. Finally, adjust the chroma gain or saturation. Again, if you're using SMPTE bars, adjust the chroma gain until the grey and blue bars are equal in intensity to the smaller bars below. Otherwise, simply adjust the chroma gain so that flesh tones suit your taste.

Because of the subjective nature of the human eye, monitor setup has been a relatively unexacting science. That is another reason why color monitors should not be consulted to determine true brightness or color values for a particular scene. When in doubt, rely on the waveform monitor and vectorscope. Recently, new advances have been made in the automatic setup and digital storage of monitor settings. Color analyzers can be attached to the front of a picture tube, and a digital readout can be obtained that describes the image being displayed by the monitor. This approach is much more objective than the human eye and tends to provide more consistent results.

At some time, you may encounter a color monitor that has an area of the screen that has a color tint, regardless of the image being displayed. This fairly common problem is caused by magnetic fields that magnetize the monitor. When you encounter this problem, it is time to **degauss** the monitor. Most studio monitors have a built-in degausser. The procedure is usually as simple as flipping a switch for a specified period of time.

DEGAUSS: To apply a random magnetic field for the purpose of erasing magnetically stored information. A videotape or a CRT can be degaussed, in the first case to erase the tape and in the second case to remove magnetic charges that cause portions of the screen to be discolored.

Projection Television

The photographic media, slides and the cinema, have long been the only suitable choice for making a presentation to a large group. Film has adequate resolution so that even when the image is projected onto a large area the result is quite satisfactory. Television, on the other hand, has always been a personal medium, one usually viewed in the home with a small audience. Only recently have attempts been made to use video as the source for group viewing. This demand on video requires a large picture with adequate image quality—not exactly a natural match. The results to date have been less than stellar. Projection televisions range in price from several hundred dollars to more than $25,000. The higher priced models clearly lead in terms of resolution and brightness.

Video Walls

A continuing trend in the field of large-scale display technology involves using video as a source of the display information. In the past, the video display technology has been limited, primarily with regard to size. Traditional CRTs have been manufactured as large as 47 inches, but these are very costly and still too small for large displays. Projection televisions can project a large image; however, the quality has often been poor, especially from angles other than straight on. **Video walls** have provided a solution. This matrix of individual monitors can be stacked atop one another to build a wall of whatever dimensions you require. The video signal is divided into the appropriate number of discrete signals and fed to the separate monitors. When viewed from a distance, the bank of monitors recreates the large picture, albeit with a grid of horizontal and vertical lines dissecting the image.

VIDEO WALLS: A display technology that uses a bank of closely spaced monitors that are individually fed parts of the complete video signal. The viewer sees one large display by looking at the monitor bank.

New Monitoring Technology

Flat panels are the wave of the future, according to many in the television display technology field. Without bulky CRTs, television monitors of the future will be only a couple of inches thick and will hang on a wall like a piece of art. Although the technology necessary for large-screen viewing is still a few years away, flat screen displays on a smaller scale are available for sale now.

Improved-definition television (IDTV) was introduced to the consumer market in 1988 and offers improved resolution from a standard NTSC signal. This is achieved by the momentary storage of a field of video (262.5 scan lines) within the monitor's digital memory. The monitor displays these lines while the following field (the interlaced 262.5 scan lines) is being displayed. In effect, IDTV monitors convert the interlaced NTSC signal into a progressively scanned NTSC signal. Although the number of scan lines is not actually increased, the apparent resolution is increased by nearly 40%.

Liquid Crystal Displays (LCDs)

The most promising technology to deliver the flat panel displays of the future is available right now in the form of liquid crystal display (LCD) monitors. These tiny portable video monitors are currently available from Sony, Hitachi, Toshiba, and Sharp. By using twisted crystal displays similar to those found in lap-top computers, LCDs make television viewing possible in a variety of situations and surroundings. Traditionally, the most serious drawback to LCD technology has been one of size. Although some had predicted an upper limit for the technology in the range of 10 to 11 inches, recent prototypes have been released with 14-inch screens and resolution of 700×550 pixels. However, these models are expected to cost much more than their picture-tube equivalents.

Sharp and other manufacturers have introduced LCD projection televisions. Combining the best attributes of LCDs—low power consumption, small size, cool operation—with the larger display area offered by projection, these units appear to be fitting the bill for many display applications.

Self-Study

Questions

1. This, the most common type of audio loudspeaker design, works like a dynamic microphone but in reverse:
   a. moving coil
   b. electrostatic
   c. woofer

2. This type of crossover system separates the audio signal into its various frequency ranges after it has been amplified:
   a. passive
   b. active
   c. limiting

3. The audio loudspeaker element responsible for reproducing the highest frequencies is the:
   a. horn
   b. woofer
   c. tweeter

4. The metal grill between the guns and the phosphor screen in a CRT is called the:
   a. underscan guard
   b. shadow mask
   c. degausser

5. This type of CRT, commonly used with S-VHS equipment, is capable of displaying a video signal with separate color and luminance signals, thus eliminating some of the cross-color, cross-luminance artifacts common to composite NTSC:
   a. delta gun
   b. Y/C
   c. RGB

6. Because the video monitor is frequently the last stop for a video signal, termination is often required at the loop-through input connectors. A switch adjacent to the BNC input connectors labeled this allows internal termination to be switched on or off:
   a. Y/C
   b. underscan
   c. Hi-Z

7. This feature, found on most broadcast-quality CRTs, permits viewing of the entire raster:
   a. blue-only switch
   b. cross-pulse
   c. underscan

8. A color monitor that displays irregularly discolored portions of the video image probably needs to be:
   a. replaced
   b. cross-pulsed
   c. degaussed

9. This switch turns off the red and green guns and is very useful when performing monitor setup procedures:
   a. blue-only
   b. underscan
   c. chroma phase

10. This new technology is expected to someday replace the cathode ray tube as the preferred video display format:
    a. video walls
    b. flat-panel LCDs
    c. two-dimensional hologram

Answers

1. a. Yes. The microphone transducer converts sound pressure into electrical signals, and the moving coil loudspeaker converts electrical signals into sound pressure. Both use a wire coil surrounded by magnets and attached to a diaphragm.
   b. No. Although this type of speaker element is sometimes used in high-quality loudspeakers, its design is based on the same principles used in the condenser microphone rather than the dynamic microphone.
   c. No. Although most woofers use the moving coil design, not all of them do.

2. a. Yes. This is generally considered to be less efficient than the active crossover, which separates the signal into frequency bands that are amplified separately.
   b. No. The active crossover system separates the signal into frequency bands before amplification. This is what makes the active crossover system so efficient.
   c. No. Try again.

3. a. No. The horn is primarily responsible for generating middle frequencies.
   b. No. On the contrary, the woofer is responsible for the reproduction of low frequencies.
   c. Yes. The tweeter, aptly named, is frequently used in a three-way loudspeaker design to reproduce frequencies above 8 kHz.

4. a. No. Try again.
   b. Yes. The shadow mask prevents the electron gun from striking phosphors of another color.
   c. No. The degausser is a feature built into many broadcast video monitors to remove from the screen the magnetic charges that can interfere with the hue of portions of the display.

5. a. No. The delta-gun CRT is defined by the arrangement of its red, green, and blue guns. No relationship exists between the gun arrangement and the type of video signal displayed.

b. Yes. The Y/C monitor is becoming popular in installations that use S-VHS and other videotape formats that record and play back a Y/C signal instead of a composite video signal. Although most Y/C monitors also accept composite video signals, there is no pronounced improvement when displaying these signals.

c. No. The RGB monitor actually accepts and displays separate red, green, and blue signals. This type of monitor, sometimes used with computers and computer graphics devices, has even less susceptibility to cross-color and cross-luminance artifacts.

6. a. No. This designation is used to distinguish a video signal that has separate luminance and chrominance components. A Y/C monitor displays the two components without the need to combine them prior to display.

b. No. The underscan switch allows the monitor to display the entire raster, making it possible for the viewer to see parts of the frame that would otherwise be hidden by the sides, top, or bottom of the raster mask.

c. Yes. Hi-Z, which stands for high impedance, indicates the switch position for the proper 75-ohm termination of a video signal. Some broadcast installations prefer to use 75-ohm terminating caps rather than the internal terminators available on most professional monitors.

7. a. No. The blue-only switch is used to turn off the red and green guns, leaving only the blue gun in operation.

b. No. Pulse cross is a feature that delays the H and V sync pulses so that the horizontal and vertical blanking intervals are visible on the monitor. The name pulse cross, sometimes referred to as pulse delay, comes from the fact that the H and V sync pulses intersect to form a cross on the screen when displayed.

c. Yes. The normal viewing mode, called overscan, actually loses part of the image beyond the side, top, and bottom edges of the screen. Underscan shrinks the display so that all of the scanning area is visible on the monitor.

8. a. No. Patches of discoloration usually do not signify a permanent problem.

b. No. See the explanation for 7b.

c. Yes. The degausser emits a random electrical charge that acts to remove lingering magnetism that sometimes builds up on the monitor screen.

9. a. Yes. Using a color bar signal as a reference, the blue-only switch makes monitor setup a less subjective procedure.

b. No. See the explanation for 6b.

c. No. Chroma phase is the name of an adjustment commonly found on TBCs.

10. a. No. Video walls use ordinary CRTs; they simply display a video signal that has been divided into smaller units so that a display of stacked monitors makes up the whole image.

b. Yes. As ATV advances hold promise for the use of increasingly larger displays, CRTs are quickly reaching the limits of their practical size. Flat-panel LCD displays promise a large viewing screen with a thin profile, which can be hung on the wall.

c. No. However, holographic images have generated their share of research over the years, and plenty of people believe that 3-D holographic displays are closer to reality than we might think.

Projects

Project 1

Survey the audio monitors in your production facilities. Classify them by the following criteria:
- type of loudspeaker elements
- number of elements per enclosure
- type of crossover network used, if any
- application of each loudspeaker, e.g., reference monitor, studio fold-back, near-field monitor

Purpose

To increase your awareness of the types of loudspeakers currently in use in your facility and to enable you to recognize the various types of loudspeaker systems.

Advice, Cautions, and Background

1. Some loudspeakers have removable grills that hide the speaker elements. Remove them carefully, and replace them when finished. If the grill is not easily removable, use the manufacturer's specification sheet to determine the specifics about the loudspeaker cabinet.

2. The crossover network will likely be hidden from view, but even if it is accessible, you would probably find it difficult to evaluate by sight. Instead, find the manufacturer's specifications for the loudspeaker system, or look for the number of amplifiers used. If there are separate amplifiers for the low, middle, and high frequencies for each loudspeaker, chances are you have an active crossover network.

How to Do the Project

1. Create a chart listing the criteria for evaluation, and fill it in with the information gathered by looking at the loudspeaker systems around your facility. Consult the manufacturer's specifications and operations manuals when necessary.

2. Turn the finished report in to your instructor.

Project 2

Perform setup procedures for a professional color monitor using color bars as a test signal.

Purpose

To learn the necessary steps to set up a color monitor and to practice the rather subjective art of matching color intensity of color bars using a blue filter or a monitor with a blue-only switch.

Advice, Cautions, and Background

1. Be sure that the monitor is properly terminated before beginning. An unterminated monitor will appear too bright, and a twice-terminated monitor will appear dark.

2. This task is much easier if you have SMPTE color bars as a source and if your monitor has a blue-only switch. If your monitor does not have a blue-only switch, a piece of Tough Blue gel (the same gel used to correct tungsten to daylight) can be used by holding it in front of your eyes as you view the monitor.

3. Without the pluge pulse found on SMPTE bars, the brightness setting may appear to be very subjective. Don't worry too much, however; monitor setup is part science and part art.

How to Do the Project:

1. Follow the steps outlined in this chapter for setting up a color monitor.

2. Have your instructor view the color bar display after you have finished.

Chapter 9
DIGITAL TECHNOLOGY

Sound and light, which are the sources of audio and video signals, exist in their natural state with values of infinitely varying degrees. Volume and pitch, color and light—they range from one extreme to the other with a complete range of values in between. Analog electronic audio and video processes attempt to ensure that this full range of values is captured, processed, and displayed. Despite its low-tech reputation, analog audio and video equipment has served broadcasters well over the years.

However, the situation is changing quickly. Increasingly, there's a good chance that the audio and video signals you experience in your home have been converted from analog to digital and back to analog again—perhaps even passing through this process several times. Somewhere between the point of origination and the point of display, the signal has been converted into a series of binary values; each note of sound and each pixel of video has been converted into its **digital** equivalent. If analog is the natural state of affairs, why is this intermediate step in the digital realm taken? In a nutshell, the intermediate step exists because digital processing is much less susceptible to noise and distortion, the very things that degrade image and sound quality. Sophisticated error correction processes available in the digital domain help to maintain the integrity of the digital signal. As long as the signal, whether it be audio or video, remains in the digital realm, it can avoid much of the deterioration familiar to analog signals. In fact, if this trend continues, it will not be long before the entire process will be digital, from the first transduction of sound or light into its electrical equivalent to the final display of digital information by loudspeaker and picture monitor. It is useful to consider the digital signal and, perhaps more important, how an analog signal is converted to digital and back again.

DIGITAL: An electronic system that processes information as a series of binary code rather than an analog waveform.

Analog to Digital

For an analog audio or video signal to be converted to a digital signal, it must undergo three steps: **sampling, quantizing,** and **coding**.

Sampling

SAMPLING: The process associated with analog-to-digital conversion in which the continuous analog waveform is divided into discrete moments in time. Also, recording a short segment of audio into digital memory to play it back as a musical element.

QUANTIZING: The process of sampling an analog signal to determine digital values equivalent to the voltage levels of the original analog signal.

CODING: The third step in the analog-to-digital process in which the information is written in binary form.

NYQUIST: Named for Harry Nyquist, the Nyquist rule states that to be able to reconstruct a sampled signal without aliases, the sampling must occur at a rate of more than twice the highest desired frequency. For example, CDs have a sampling rate of 44.1 kHz and allow signals up to 20 kHz to be recorded.

Sampling can be described as taking a reading of the signal, or measuring its amplitude, at a particular moment in time—taking its picture, if you will. Taking the motion picture analogy a little farther, the sampling rate determines how often the picture or reading is taken. Motion pictures are typically captured on film with a frame rate of 24 frames per second. Every 1/24 second, a new image is captured. If the frame rate were slower, you would perceive jittery motion, i.e., the motion wouldn't appear to be fluid or you would perceive false motion. An example of this is the film illusion of wagon wheels that appear to rotate in the wrong direction. If, on the other hand, the sampling rate were higher than 24 fps, you would perceive even greater detail in motion sequences. However, the economics of film stock and processing determined that 24 fps is sufficient for an acceptable reproduction of motion. Understandably, the sampling rate for analog to digital conversion of an audio or video signal is much higher than film's 24 fps. According to the **Nyquist rule**, the optimum sampling rate for analog to digital conversion is at least twice the highest frequency of the signal you wish to sample. For audio, which has a maximum frequency range or bandwidth of about 20 kHz, a sampling frequency of 44.1 kHz will provide the necessary definition. Because 44.1 kHz is the sampling rate for CDs, it has come to be described as *CD-quality* digital audio. At the low end of the audio spectrum, digital audio for commercial telephone is sampled at 8 kHz, providing for a bandwidth of 3.2 kHz.

Quantizing

Quantizing, the second step in the analog to digital process, has to do with assigning a numerical value to the analog level. If sampling can be described as a time-discretionary process, quantizing is an amplitude-discretionary one. In a single-bit **binary** system, there are two levels: on and off. In the binary system, the number of **bits** used to make up the digital word determines the number of steps or levels. Using the formula 2^x, where x = the number of bits, provides the number of steps or levels. With a two-bit system, there are four: two to the power of two. A three-bit system allows for eight levels, four bits allows for 16, five for 32, six for 64, seven for 128, and an eight-bit system has 256. Normally, an eight-bit word is also known as a **byte**. A 16-bit system, like that used for CD audio, has 65,536 discrete levels or possible values. Of course, the more levels, the more faithful the quantizing process is to the true analog value. Anytime that the actual value falls between steps or levels, which is often, quantizing error is possible. The more levels, the smaller the steps and the less noticeable the error. In digital audio processing, the dynamic range and the number of bits are directly related. Each additional bit improves the signal-to-noise ratio by 6 dB. Again, using CD audio as a reference, the 16-bit quantizing for CD audio provides a S/N ratio of about 98 dB.

Aliasing

The analog-to-digital (A/D) and digital-to-analog (D/A) process is not without its weaknesses. Because an analog signal is infinitely variable, it is impossible to create a perfect digital equivalent. It may be good enough, more accurate than human eyes and ears can discern but imperfect nonetheless. However, when these errors are great enough to cause visible or audible distortion, they are commonly known as **aliasing**. Aliasing may be evident in the audio digitizing process as whistling and in video as moiré effect.

Digital Audio

By far the most significant advance in audio production and post-production in recent years has resulted from the proliferation of digital recording. Now that the audio signal can be digitally recorded onto magnetic tape or disk, it is a simple matter to manipulate and alter the audio signal in the digital realm. Although digital storage on disk provides the advantage of random access, floppy disks and hard disks have limited capacity. Magnetic audio tape not only provides greater capacity but makes storage expansion as simple as loading another reel of tape. Professional digital audio tape recording formats include DAT, digital audio stationary head (DASH), Pro-Digi, and the 3/4-inch VCR-

BINARY: Mathematical representation of a number to base 2, i.e., with only two states: 1 and 0; on and off; or high and low. Requires a greater number of digits than base 10, e.g., 254 = 1111110.

BIT: One digit of binary information. A contraction of the term *binary digit*, it has a value of either 0 or 1.

BYTE: One byte of digital video information is a packet of bits, usually but not always eight. In the digital video domain, a byte is used to represent the luminance or chrominance level. One thousand bytes is one kilobyte (kB) and one million bytes is one megabyte (MB).

ALIASING: Undesirable "beating" effects caused by sampling frequencies that are too low to faithfully reproduce image detail. Aliasing is evidenced in video as jaggies, moiré or video noise, and in audio as unwanted frequencies that result when filtering is not used to eliminate frequencies above the range of the sampling frequency.

based PCM format. Another digital audio recording format, digital compact cassette (DCC), was introduced recently. Philips presented the new digital format at the 1991 Consumer Electronic Show and hopes to have production units available in 1992. DCC units, which use the same size cassette shell, a stationary head, and the same tape speed as the analog compact cassette, will reportedly be able to play back both standard audio cassettes and the new digitally recorded cassettes. It's still too early to tell if DCC will enjoy the same level of acceptance by industry and broadcast professionals as the DAT format has.

What's DAT?

The big news in digital audio in the consumer and professional arenas is the availability of DAT recorders. DAT's sound quality is nearly identical to that available on compact disc, but has the advantage of being a recordable format. With portable units weighing in at less than three pounds, cassette tape half the size of analog compact cassettes, and a record time of 2 hours per cassette, DAT holds great promise for a variety of recording applications. Fast search speeds allow any point on a 2-hour DAT cassette to be located in less than 30 seconds. DAT is available in two formats: R-DAT and S-DAT. R-DAT uses a rotating head drum configuration, and S-DAT uses a stationary head. Yamaha may be the only commercial company taking the S-DAT approach. R-DAT equipment, however, is being marketed by numerous consumer and broadcast equipment manufacturers. Consumer and professional DAT machines operate at one or more of three sampling frequencies: 48 kHz, 44.1 kHz, and 32 kHz. The standard sampling rate is 48 kHz; 44.1 kHz permits compatibility with the CD digital recording format; and 32 kHz allows longer recording time at slightly lower fidelity. Most consumer and professional DAT recorders will automatically play back tapes recorded at any of the three sampling rates.

The marketing of DAT recorders to consumers has cleared one of the major legal hurdles set up by the Record Industry Association of America (RIAA), the agency that represents the artists and musicians who stand to lose from widespread illegal digital copying by consumers. The RIAA's acceptance of DAT hardware was achieved by the inclusion of serial copy management system (SCMS) technology. SCMS allows DAT recording of a digital source, such as a CD, but the resultant recording cannot be used to make a subsequent recording. DAT recorders are becoming widely available on the consumer and broadcast equipment markets.

It is important to note that DAT, although first and foremost a consumer format, is being embraced by the professional broadcasting community and has been used professionally for some time. Even the film production community has found a use for portable DAT recorders. Because the DAT recorder always records and plays back at perfect speed (due to its internal sampling rate of 48 kHz), the portable DAT recorder can be used to record location audio for film. DAT's other advantages over the industry-standard Nagra 1/4-inch reel-to-reel ATR include all of the advantages of digital recording and its compact size: a full day's worth of recording stock can fit in your shirt pocket. DAT obviously has widespread application for the audio, video, and film business, and with the addition of time code, the DAT format has almost ensured its future in the broadcast industry.

Digital Disk Recording

Digital tape recording is being used widely in multitrack production and in the premastering stages of production. In audio post-production, however, the story is a little different. Digital disk recording systems, although limited in storage capacity, permit random access to the audio material. Random access makes the editing process incredibly faster and more efficient, as you'll recall from Chapter 7. However, there is a price to pay. Digitized audio can use up a lot of disk space quickly. At a sampling rate of 44.1 kHz, a mono signal consumes 88.2 kilobytes of disk space each second, or 5.29 megabytes (MB) each minute. To address the storage problem, large computer hard disks are connected to achieve several gigabytes of storage; some systems provide up to 2 hours of CD-quality audio at a sampling rate of 44.1 MHz. Another format allows digital recording and random access but is limited in another way. Write once read many (**WORM**) optical disks permit larger amounts of storage, but like its name implies, the WORM disk can only be written to, or recorded, once. This makes WORM technology most useful when completed projects need to be stored for later retrieval. Many expect the arrival of erasable/recordable optical disks to alleviate the digital storage capacity problems for digital audio post-production applications.

WORM: Write once read many. A type of optical disk recording that was popularized by the success of audio disks (CDs) and video disks (laserdiscs). WORMs, as the name implies, can only be recorded once. This makes them suitable for permanent storage but limits their use in a post-production or creative environment.

One of the great advantages of digital audio editing on a disk-based workstation is its nondestructive nature. Analog audio editing has for decades involved cutting and pasting, physically moving segments in relation to other segments. To change an analog edit, you must go back and physically perform another or several other edits. In the digital editing process, editing instructions are stored in an edit list, much like the EDL associated with analog videotape editing. When changes are made, the EDL is simply updated to reflect the change, and there is no need to actually reassemble program material on the hard disk. During the entire process, the integrity of the original source material is preserved. The data remain unchanged on the disk. In fact, the material does not actually exist in its edited form until the finished program is dumped from the disk to another recording medium, i.e., analog or digital tape. Say "goodbye" to your razor blades and grease pencils: digital audio editing is here to stay.

Digital Video

Digital video, like digital audio, has inherent advantages over analog video for processing and recording applications. One obvious advantage is its freedom from generational loss. Once a signal is in the digital realm, duplication of the signal introduces minimal signal loss, especially when compared to analog processing and duplication. Another advantage of digital processing is the ability to assign absolute values to variable parameters, thus making them recallable and repeatable. Once a video signal is converted from analog to digital, any number of maneuvers can be performed to create dazzling special effects. Digital video effects are simply manipulations of a digital video signal. Many experts believe that the day is not far off when every step of the production process, from origination to display, will be performed in the digital realm. To understand the digital video signal, you should consider the same steps explained earlier in the analog-to-digital conversion process.

Sampling and Quantizing Specifications for Digital Video

As you already know, the frequency and bandwidth of the video signal is much greater than that of the audio signal. Sampling the analog video signal at too low a rate can introduce undesirable side effects known as *aliasing*, an important topic that is considered later in this chapter. To minimize the effects of aliasing, composite video signals are sampled at approximately four times the NTSC subcarrier frequency (4 × fSC) of 3.58 MHz, or 14.318 MHz.

Comité Consultatif International des Radio-communications (**CCIR**) **Recommendation 601**, an international standard for digital component video, specifies a sampling rate expressed as a ratio of 4:2:2. That is, the luminance signal is sampled at four times the base of 3.375 MHz, and the two color-difference signals are sampled at two times the base frequency. Although a sampling rate of four times subcarrier provides excellent results, available recording and transmission equipment dictates that care must be taken to minimize the bandwidth of the digital video signal. The sampling rates for 4:2:2 video are as follows:

Y (luminance)	sampled at 13.5 MHz
R-Y (Cr)	sampled at 6.75 MHz
B-Y (Cb)	sampled at 6.75 MHz

In situations where signal quality is paramount and signal bandwidth concerns are minimal, digital video sampling rates take on even greater proportions. Sampling frequencies used for high-end digital component video processing include **4:4:4** and **4:4:4:4** (also known as *full-bandwidth component digital video*). No practical device presently exists to record the 4:4:4:4 signal (although some post-production houses are using a pair of D-1 VTRs to record the 4×4 signal, and two disk recorders from one manufacturer can be chained together to achieve the same result). Regardless, some manufacturers, in their quest for optimum signal quality, are building 4:4:4:4 processing into their equipment. Broadcast equipment using 4:4:4:4 processing includes Ampex Alex and Abekas A72 character generators, ColorGraphics DP-422 and Digital F/X Composium graphics systems, and Quantel's Harry. These manufacturers expect 4:4:4:4 processing to be widely adopted in the near future and to serve as a bridge to future HDTV developments.

With regard to quantizing, digital video actually deals in fewer levels than the digital audio system described earlier. CCIR Recommendation 601, also known as **4:2:2**, stipulates eight bits for coding. Eight bits, as you may recall, allows 256 possible combinations—in this case, 256 shades of grey. Tests have shown that the average viewer cannot differentiate between more than 200 shades of grey, so a system capable of 256 shades of grey should theoretically be as good as anyone will ever need. However, some digital video devices use nine- or ten-bit systems. These may or may not provide a better image than the eight-bit system. Ultimately, it may come down to a matter of subjective preference.

CCIR RECOMMENDATION 601: Encoding parameters of digital television for studios. The international standard for digitizing component color television video in both 525- and 625-line systems, for which it defines 4:2:2 sampling at 13.5 MHz with 720 luminance samples per active line and eight-bit digitizing.

4:4:4: One of the ratios of sampling frequencies used to digitize the luminance and color-difference components (Y, Cb, Cr) or the RGB components of component video. It differs from 4:2:2 in that all components are sampled at 13.5 MHz.

4:4:4:4: The same as 4:4:4, except that the key signal is included as a fourth component, also sampled at 13.5 MHz.

4:2:2: One of the ratios of sampling frequencies used to digitize the luminance and color difference components (Y, Cb, Cr) of component video. The luminance signal is sampled at 13.5 MHz (4 times 3.37 MHz), and the two color-difference signals (Cb, Cr) are sampled at 6.75 MHz (2 times 3.37 MHz).

Video Aliasing

Aliasing, as you recall, is a result of imperfections in the analog-to-digital conversion process. Perhaps the most graphic example of video aliasing is seen in the crude characters generated by very simple character generators, e.g., the type built into consumer camcorders. You may have noticed the stair-stepping or jagged edges on what would normally be rounded characters. Aliasing can be reduced or masked by sophisticated mathematical operations. These mathematical formulas perform rounding off, or smoothing, routines that are known as **anti-aliasing**. Almost all broadcast-quality character generators contain anti-aliasing circuitry to reduce these unwanted artifacts.

Digital Videotape Recorders

The professional videotape recording formats currently available that use digital recording technology are D-1, D-2, and D-3. Developed by Sony, Ampex, and Panasonic, respectively, these formats differ in many ways. Most notably the D-1 format is a component digital recorder, and the D-2 and D-3 formats digitally record a composite video signal. All of the digital VTRs are available in a studio configuration, while the D-2 and D-3 formats are also available as portable units. The acceptance of digital videotape recorders by broadcasters and others in the video production industry has been mixed. Broadcasters have not rushed to purchase the D-1 format machines, largely due to the high cost and the need for component signal processing to see the full benefit from this digital format. However, the D-1 format is making inroads into high-end post-production facilities and computer graphics houses. On the other hand, D-2 has seen considerable acceptance in a wide range of facilities and applications. Because the D-2 format is composite, it plugs into an existing facility without any rewiring or other preparation. As existing 1-inch machines approach the end of their useful lives, many broadcasters consider the D-2 machines to be the logical replacement.

Although the manufacturers of digital videotape recorders would like people to think that they have invented the perfect recording device (they even go so far as to call digital copies *clones* instead of *dubs*), they are stretching the truth just a bit. Digital VTRs (DVTRs), although free from the signal degradation and noise inherent in analog reproduction, are still not perfect in terms of generational performance. Unlike hard-disk and solid-state recorders, DVTRs must work with the medium of magnetic tape, which is subject to dropout, oxide wear, stretching, and associated problems. Error correction and error concealment circuitry can fix or mask most of the problems introduced by recording on magnetic tape; however, it is possible to introduce texturing and other artifacts. In normal operation, DVTRs should be capable of at least 20 generations before any objectionable degradation of the signal occurs.

The trend toward digital has made analog VTR manufacturers stand up and take notice. Multigenerational performance is one of digital's most important attributes and one area where it would appear that analog VTRs can not compete. However, at trade shows in 1989, Ampex demonstrated the 1-inch model VPR-3 with their Zeus processor. The demonstration included a split screen between a seventh and twenty-third generation dub. According to those who compared the results, 1-inch type C with the right equipment and care can produce excellent multigenerational results. Although it is true that under controlled circumstances 1-inch type C recorders can make multiple generations with little loss of quality, a great deal depends on operator setup.

D-1, D-2, D-3...

In 1981, the SMPTE and the EBU performed joint digital video tests that eventually resulted in the first worldwide digital video standard. The standard was acknowledged by the CCIR as Recommendation 601. The first digital video recorder to employ Recommendation 601 came to be known as the D-1 VTR. The D-1 machine can record NTSC, PAL, or SECAM. Note, however, that it is not a conversion device; it simply records the incoming signal and plays it back in the same system. It uses the component digital standard and is considered to be top-of-the-line in terms of recording quality and transparency. However, the D-1 format requires the use of external peripheral devices compatible with the component signal. Also, the D-1 format has limited slow-motion capabilities. Many of the first D-1 machines were delivered to high-end post-production houses such as Charlex, Optimus, and Grace & Wilde. Because D-1 requires component video signals to realize its full potential, it has not become a popular format for traditional broadcast installations. The videotape cassette used for D-1 comes in three housing sizes: small, with 11 minutes running time; medium, with 34 minutes; and large, with either 76 or 94 minutes. The tape is 19mm wide, approximately 3/4 inch, and uses a high-energy oxide, nonmetal formulation.

D-2 is a more recent addition to the already teeming waters of video recording formats. When Ampex announced the D-2 format in 1986, some broadcasters were outraged that a manufacturer would so quickly challenge the relatively new D-1 worldwide digital standard. Ampex was accused of many things, including attempting to sabotage Sony's new D-1 machine. However, Sony decided to join Ampex in bringing the D-2 machine to market, and much has changed since then. Early D-2 VTRs have had their share of reliability problems, but second generation machines appear to have resolved most of them. Today the D-2 format displays some amazingly attractive attributes, including these:

- D-2 maintains nearly perfect quality for at least 20 generations of dubbing and editing—all this in the NTSC composite format.
- D-2, because it is composite and NTSC, can plug into existing NTSC systems without the triple routing and complex additional equipment required by the component digital standard.
- D-2 has four channels of digital audio (with quality even better than CD), a separate channel for time code, and another for cuing.
  NOTE: Initial problems locating audio edit points in slow or still speeds have apparently been remedied by Ampex's Accu-Mark™ audio recovery process.
- D-2 can record nearly 3.5 hours on one cassette.
- D-2 allows you to see color pictures with shuttle in either direction at ±60 times normal play speed. (Sony's new DVR-28 allows for 100X shuttle)
- D-2 has variable speed motion from −1 to +3 times normal play speed.
- D-2 will play back perfect pictures even if the tape is scratched or two heads become clogged or broken.

- D-2 can mistrack by 50% in either direction and still offer perfect pictures and sound.
- D-2 is priced about the same as a top-of-the-line 1-inch type C VTR.
- D-2 cassettes use metal-particle tape, are 19mm wide (about 3/4 inch) and come in lengths of 32, 94, and 208 minutes. Ampex is marketing a multicartridge recorder/player (the ACR-225) that is based on the D-2 format. Now television station automation can take advantage of digital video recording technology.

Panasonic's D-3 format is the most recent addition to the digital video recording arsenal, with prototypes introduced at the 1990 NAB convention in Atlanta. Unlike the D-1 and D-2 standards which use 19mm tape, the D-3 format uses 1/2-inch tape. The smaller videocassette reduces storage space and makes a digital camcorder a possibility. D-3 has received the SMPTE's approval as a new recording standard and has been adopted by several major users, including the BBC. Another important development in favor of D-3's acceptance is the decision by the Comite Organizador Olimpico Barcelona 1992 to designate it as official broadcast equipment for the 1992 Barcelona Olympic Games. Panasonic's parent company, Matsushita expects to supply approximately 400 studio VCRs and camcorders. The camcorder model will be paired with Panasonic's digital camera for a completely digital one-piece unit.

Disk-Based Recording Systems

There is quite a bit of interest in disk-based recording systems for several reasons, the most important being the random-access nature of disk-based recording versus the linear nature of tape. Dedicated disk recorders such as the Abekas A-64 are extremely popular in high-end post-production facilities, especially those that work in computer graphics. The drawback to disk recording has been the short record durations available. Currently, an A-64 equipped with 2.64 gigabytes (GB) of hard-disk storage can record 100 seconds of full-motion, digital video. This is adequate for commercial work, but further advances in hard disk and optical disk storage will be necessary for other applications. Some companies are using optical disk recording (WORM drives) and playback systems for off-line editing. Others are using PCs and hard disks to record compressed video for the same purpose. The future of disk-based video recording is promising for both the high-end (large post houses) and the low-end (desktop video) of video production.

Dynamic Random-Access Memory (DRAM)

Dynamic random-access memory (DRAM) is used for the storage of video still images. Framestores, TBCs, and still store devices use DRAM chips for the storage of single frames of video. The advantages of DRAM storage are almost instantaneous recall and display of a great number of images and the ability to instantly update the list of images. DRAM chips are available in 256 kilobits (kb) and 1 Mb per integrated circuit. The use of DRAM on a larger scale is very attractive because, unlike other types of digital storage, it does not depend on mechanical processes to store and retrieve information. However, one disadvantage of DRAM is that it is volatile storage, i.e., memory is lost when power is removed.

Harry Who?

Quantel's Harry is a digital paint box, hard-disk video recorder, edit controller, multitrack digital audio mixer, and video keyer combined into one unit. It provides high-end manipulation of the video and audio signal in real time for computer video graphics and post-production. This is the first integrated digital audio and video processor available in one package from a manufacturer. Harry's internal memory is provided by Winchester hard disks, but the ability to interface with digital VTRs increases potential storage capability.

Computers

The radio, television, and film industries cannot ignore computers, even if they wished to for some reason. Computer technology has so deeply infiltrated all levels of production and post-production that it is no longer a question of *whether* computers will be fully integrated into all phases of the production process but *how quickly* it will happen. At first glance, the connection between the broadcast media and the computer sciences may not be readily apparent. Although both businesses rely heavily on hardware and software, the broadcast media have been primarily concerned with pictures and sound, and computers have historically been most useful crunching numbers. What is interesting, however, is the way in which the computer industry is discovering video and sound and the way that broadcasters have integrated computer technology in almost every area of production. Computers are becoming such a large part of the broadcast production business that no one working in the business can afford to be ignorant of their use and operation.

The presence of computers almost everywhere may be accurately construed to suggest their importance to industry in general and to the mass media in particular. The computerized news room is an example of the way that computers are being integrated within the television industry in a very obvious manner. However, in many broadcast applications, what may actually be less obvious is the computers themselves. Certainly, the ubiquitous PC is showing up more frequently than ever in the production environment. More often than not, however, the computer technology is being included in hardware that bears remarkably little resemblance to a computer. Computer-controlled camera setup, computerized editing, computer-controlled automation of cameras, VTRs, audio consoles, etc., and computerized graphics production and signal processing are already commonplace. Microprocessors are showing up in VTRs, audio mixers and processors, studio cameras and transmission equipment—the list goes on and on. The new D-2 format VTRs contain the computing power of a personal computer, but the operator is only aware of enhanced operating modes. The operator does not interface with a computer but with a very "smart" VTR.

Perhaps your interest in the audio/visual media does not include preoccupation with the latest hardware toys and their computerized bells and whistles. Even if your interests are limited to purely creative endeavors, for example, writing for radio or television or film, remember that word processing and scripting software has almost made paper and pencil go the way of the quill pen. Today, it can safely be said that any aspiring broadcast professional who is not at least marginally comfortable with a computer keyboard and CRT is already at a disadvantage in this business. It would be beneficial to look first at the many ways in which computers and computer technology are currently integrated into the radio and television production process and then to look at the direction in which things appear to be headed.

Computer Technology Today

Television cameras sometimes feature computer setup, i.e., parameters of white and black balance, registration, and timing are stored in computer memory for instant recall on demand. Panasonic recently introduced the AQ-20, which uses all digital circuitry for processing and setup.

Videotape recorders, including the professional 1/2-inch, 1-inch and digital formats, use extensive microprocessor control for tape transport and editing features. Like cameras, the recall of VTR setup parameters is another feature made possible by computer memory. As cameras and VTRs incorporate more digital circuitry, the storage and recall of settings and control parameters will increase.

Another piece of equipment that makes use of digital and microprocessor technology is the TBC. Whether it takes the form of an outboard box or a plug-in circuit board, the TBC relies heavily on digital memory. Some outboard models feature two-dimensional digital video effects, and others allow the proc amp setting to be memorized for later recall. Both of these features exist courtesy of computer technology.

Today's digital switchers incorporate microprocessors to set levels and recall parameters for keys and wipes. Computers in switchers talk to other computers, e.g., computer editors and digital effects devices.

Digital video effects systems are little more than computers with video A/D and D/A converters. NEC's Digital Video Effects (DVE®), Ampex's ADO®, and Quantel's Mirage are just a few examples. Once the video is digitized, mathematical formulas are applied to create effects ranging from simple compression to complicated three-dimensional flips and turns.

Computerized editing has been the standard for video post-production for more than a decade. Computer-based edit controllers, such as those bearing the CMX label, are the industry standard. These computerized edit controllers use either **RS-232** or **RS-422** protocol to communicate with other microprocessor-controlled, or *smart* devices, e.g., switchers, videotape machines, and DVE devices.

RS-232: Recommended Standard 232. A communications protocol popular for serial communication between computer devices.

RS-422: Recommended Standard 422. A communications protocol popular in the video post-production environment. Computerized editors and VTRs commonly use RS-422 interfaces.

Computer graphics, including character generators and paint systems, may be the most obvious domain of computers in the broadcasting environment. Dedicated systems by Quantel, Dubner, Wavefront, and Iris can produce animated three-dimensional graphics for a "network-quality" look. Computer graphics is also an area where low-end computers, i.e., PCs, have been able to make the greatest inroads. Three-dimensional systems are available for the IBM PCs, and the Commodore Amiga offers sophisticated graphics at a budget reasonable for even the smallest station or production company.

In the area of audio production, computers are becoming standard equipment. Audio post-production is becoming increasingly computerized with the use of MIDI technology, samplers, digital audio workstations, and digital recording.

175

Another important technology at the local station level is automation. Both radio and television are taking advantage of the technology to reduce human participation in areas that can be handled by computers and robotic machinery. Robotic cameras, automated audio tape and videotape playback systems, and automated consoles are finding acceptance in the smallest and largest of radio and television stations. And the list of jobs falling to computers and robotic technology is growing every day.

Another area where computers are being used extensively in the radio, television, and film industries is in the area of budgeting. Producers and production managers prepare commercial bids and budget proposals and perform production budget tracking using PCs and spreadsheet software. Custom software has been developed for the film and television industry to facilitate budgeting for location and studio production.

Finally, script preparation, as with any writing task, is now faster and more efficient with the use of a personal computer and word-processing software—especially when scripts must go through the approval and revision process. Most programs allow both the traditional screenplay format and the two-column audiovisual (A/V) format that is used in television news and feature scripting. Storyboarding is also possible with some software packages. Scripting software is available that allows the finished script to be displayed directly as prompter copy, thus eliminating the intermediate step of printing out hard copy.

To increase your computer literacy, it would be helpful to consider some basic principles of computers and computer technology. One place to start is with the terms *hardware* and *software*.

Hardware

CPU: Central processing unit. This is the collective name for the primary memory, logic units, and the main control unit of a computer system.

Hardware is essentially the nuts and bolts of the computer—or, in this case, the silicon chips and integrated circuits. Don't confuse *silicon* (this word is derived from the root word for *sand* or *glass*) with *silicone* (a type of plastic used for caulking and plastic surgery). As sophisticated as computers may appear to be, the fact remains that they are simply electronic devices that are capable of extremely fast calculations—number crunching. Computer hardware may be specified in terms of its central processing unit (**CPU**), memory, input and output (I/O), and mass storage devices. When you understand the hardware related to personal computers, you'll be better able to understand the computers that are being used in the broadcasting environment.

The Central Processing Unit

In the IBM family of PCs, you'll find microchips manufactured by the Intel Corporation. The early IBM PCs used the 8088 chip; since then they have incorporated faster and more powerful chips, including the 80286, 80386, and 80486. Chips used in computer CPUs are rated according to clock speed, which in this line of chips ranges from 4.7 to 33 MHz and higher. The higher the clock speed, the faster the computer can process information and make calculations. However, clock speed is only part of the equation. The bit size of the processor is also very important in determining the speed with which the microprocessor can execute instructions. An 8088 chip rated at 10 MHz is not as fast

176

as an 80286 chip rated at 10 MHz. The difference is that the 8088 chip has an eight-bit bus, and the 80286 chip has a 16-bit bus, which permits much higher performance. The same holds true for 386 and 486 chips.

The Apple Macintosh and Commodore Amiga use the Motorola family of microchips, including the 68000, 68020, 68030, and the newest member, the 68040 chip. These operate at clock speeds ranging from 10 to 40 MHz. A new chip by Motorola, the 96002 floating-point dual-port processor operating in either the 32- or 44–bit mode, has the potential for amazing graphics and sound manipulation for digital audio and video workstations. Motorola refers to this chip as the *Media Engine* and sees applications in video compression, computer animation, and facsimile transmissions, just to name a few.

Memory

Two types of memory are found in most computers: **read-only memory (ROM)** and **random-access memory (RAM)**. Read-only memory holds information in memory where it may be utilized by the computer. Unlike other types of memory, ROM is usually permanent, i.e., it is not lost when power is turned off. Operating systems are sometimes stored in a type of ROM known as *programmable read-only memory (PROM)*. PROM chips that contain operating instructions allow nearly instantaneous start-up. Because there is no delay to allow the computer to access information stored on magnetic media, the log-on or start-up procedure is very quickly executed. Another type of ROM is known as *erasable programmable read-only memory (EPROM)*.

Random-access memory, either static RAM or dynamic RAM (DRAM), is volatile memory, i.e., if for any reason the power is lost or interrupted, the memory is erased. When the computer's main power is turned off, static RAM memory is lost unless there is a battery back-up. DRAM, on the other hand, must be refreshed by the CPU every few milliseconds. Because of this fact, it cannot be preserved by battery back-up. However, because DRAM has a higher memory density per chip, it is often used in computers that require a large amount of RAM. Of course both types of RAM can easily be stored permanently by recording onto magnetic media such as disk or tape.

ROM: Read-only memory. A type of primary storage written to only once, after which it can be read from but not altered.

RAM: Random-access memory. A type of primary storage in which any randomly selected segment can be accessed at will for reading or writing.

I/O Devices

Input and output devices are used to get information into and out of the computer. Typical input devices include the keyboard, digitizing camera, scanner, mouse, and light pen. Output devices include the printer, video display (CRT), or NTSC video to be displayed or recorded by a videotape recorder.

Mass Storage Devices

The ability to record information in its digital form allows information and programs to be shared between compatible computers or stored for any length of time. The most common type of storage today, the magnetic floppy disk, is available in a variety of formats. Two of the more popular physical sizes are the 5.25-inch mini-floppy disk and the 3.5-inch micro-floppy disk. Memory size varies from 360 kilobytes (kB) to 1.44

MB. Some of the formats are 5.25-inch/360 kB, 5.25-inch/1.2 MB, 3.5-inch/720 kB, 3.5-inch/800 kB, and 3.5-inch/1.44 MB. Some of the older computer editors used 8-inch floppy disks, but these are becoming extremely rare.

HARD DISK: A device for magnetic storage of digital information. Also known as *fixed disks* or *Winchester disks*, hard disks record information on a spinning disk of smooth metal, usually aluminum. Hard disks have faster access time and contain a greater amount of storage than floppy disks.

Magnetic fixed disks, also known as **hard disks** or *Winchester disks*, are not only extremely popular, many consider them to be a necessity. Hard disks cover quite a range in storage capacity, from 20 to more than 600 MB. Hard disks with large capacities are especially important when working with digitized audio or video because digital files can require large amounts of storage space. In fact, some digital audio and video devices chain together large hard disks to achieve several gigabytes of storage capacity. One example is the Avid Media Composer, a random-access off-line editing system built around the Macintosh computer. It uses seven 600-MB hard disks to provide storage of digitized audio and video.

Software

Software is what makes a computer useful. Software, or computer programs, permit the operator to access the power of the computer using simple commands or instructions. Most computer programs for personal computers are sold separately from the computer hardware and are made to work with one particular type or configuration of computer system. However, many dedicated computer systems used in radio, television, or film production have custom software designed to achieve a very specific task. The computer program for the Chyron Scribe character generator is dramatically different from the program used to create graphic images in the Quantel Paintbox®, which is very different from the software used to create 3-D graphics using the Wavefront system. The same is true with digital audio workstation software and computer editing systems. These hardware manufacturers must create proprietary software, and this can be an extremely costly proposition.

On the other hand, software is fairly inexpensive when it is marketed to a large user base. This is the case with software developed for the millions of personal computers in use today. However, this software is generally much less powerful at performing the one specific task for which dedicated computer systems are designed. It is versatility that the PC achieves so successfully. The same computer, with several software packages, could effectively be used to write a script, create 2-D and even 3-D graphics, perform off-line editing functions, and invoice the client. Software for personal computers is either command based, such as MS-DOS for the IBM family of PCs, or it can be graphically based, such as the operating systems for the Apple Macintosh, Commodore Amiga, and the NeXT™ computer. Graphically based software is generally perceived to be easier to learn and more quickly embraced by the graphically oriented media businesses. With the release of Microsoft Windows, the DOS family of personal computers is becoming much more graphically oriented. Like the dedicated hardware/software systems mentioned earlier, programs designed for PCs are machine specific. Software written for an IBM or compatible machine will not work on an Apple Macintosh computer or vice versa, although quite frequently a software developer releases versions compatible with more than one type of PC hardware. Recent agreements between Apple and IBM may be the start of a trend toward more compatibility between operating systems and software.

Communications Protocol

One of the wonderful things about computers is their ability to communicate with other computers. Communications between computers may mean simply that they are sharing information or data, or it may involve one computer actually controlling the operation of the other. In either case, for successful communication to take place, they must first speak the same language. A few of the languages spoken between computers in the broadcasting arena are RS-232, RS-422, MIDI, and SMPTE time code.

One way for computers to share information is to be physically connected by a multipin cable. For serial communications over multipin cable, RS-232 is the recommended standard. In the television production arena, RS-422 is being used by an increasing number of computer editors to speak to VTRs, switchers, and digital video devices.

Musical instrument digital interface (**MIDI**) is a standardized protocol for computer control of electronic musical instruments. MIDI allows a composer or musician to sequence a number of electronic musical instruments to play back together, thus providing the equivalent of individual control of a whole orchestra or band. MIDI technology is used extensively in the production of audio for radio, television, and film.

MIDI: Musical instrument digital interface. This is the standard protocol for computers to interface with musical instruments (synthesizers) to permit sequential playback.

SMPTE time code is actually an analog audio data signal used in the process of audio and video post-production. For a more complete discussion of SMPTE time code, see "Time Code" in Chapter 7.

Computer Technology at Work in the Industry

Station Automation

Radio has for some time had available the necessary technology to format a local station and provide programming with limited human facilitation. Programming delivered via satellite and automated tape playback machines for locally inserted spots, news, and weather have allowed stations to operate with only a handful of people on staff, and these are primarily in sales and management. Television is not far behind with the deployment of the computer-controlled cart machine, which selects the scheduled programs and spots from the library, inserts them into the VTRs, plays them at the appropriate time, logs when each spot was run, and prepares the invoice for billing. If all of this is beginning to make you want to reconsider a career in broadcasting, take heart—there are still a few jobs that computers have not been able to master . . . yet!

The Computerized Newsroom

Broadcast news production, to a degree unlike any other style of production, is a slave to the clock. The task of getting a story on the air involves drawing information from various sources, compiling and editing the information, having it cleared by appropriate members of management, and presenting it in an appropriate form to be read by on-air

LAN: Local area network. A data communications system consisting of two or more microcomputers physically connected together with some type of wire or cable, over which data is transmitted.

MODEM: Modulator/demodulator. This device converts a signal from a computer to one capable of being sent over telephone lines. A modem at the other end converts the signal back to one able to be received by another computer.

talent. Computers have simplified and streamlined this process tremendously. Instead of wire copy being ripped from the printer and typed on a typewriter, computer files are downloaded from the wire services and edited within the computer. When the story is finished, it is transferred by **local area network** (**LAN**) or **modem** (modulator/demodulator) to the news editor's PC. The editor makes changes, offers suggestions, and signs off on the story. From there, the file is sent to the TelePrompTer® device and displayed in front of the camera lens for the talent. At no time in the entire process is it necessary for the story to be printed out on paper. Instead of creating and editing hard copy, the story originates and concludes as electronic characters. Currently, two of the more popular newsroom computer systems are marketed by Dynatech (NewStar) and BASYS Automation Systems Incorporated.

There is considerable interest in the integration of the newsroom computer system with the automated news studio for an automated newscast. In this scenario, the only personnel required would be the on-air talent, a news producer, and a technician to program the news show prior to air time. All other functions, including camera work, switching, character generation, and VTR playback, would be automated and controlled by computers. Perhaps the next question to ask is when does the on-air talent become expendable?

Computers and Celluloid

Francis Coppola is considered by many to be a pioneer in the concept of the electronic cinema. The idea actually involves almost every aspect of a feature film's production, from scripting to post-production. The principal steps are scripting (word processing), storyboarding (computer graphics), video rehearsal (computerized graphic animation), film-to-video transfers (edge numbering technology and software), nonlinear editing (disk-based recording), and audio sweetening (MIDI technology). Some believe that once all of these steps are brought under the control of one computerized workstation, the flexibility, cost efficiency, and personal control gained will make it a very attractive option for the filmmaker.

Personal Computers in Production

The integration of computers in the radio, television, and film industries has been phenomenal. One reason for this has been the invention and refinement of the personal computer. The PC has become a powerful tool for all but the most demanding computing tasks. The PCs of today have more memory and power than mainframe computers of just a few years ago. Not only are the computers becoming more powerful, they are at the same time becoming more user-friendly. It is only natural that this powerful tool will be put to use in more applications in the coming years. Following are the three most popular personal computers and some notes regarding their application in today's market.

IBM and the IBM-compatible PC: The mainstay of the business community; traditional, ubiquitous, and the clones are inexpensive.

Apple Macintosh: User-friendly, graphically oriented, desktop publishing and MIDI strengths; costly, and slow to get into video applications, although the new line of Macs has expanded built-in video capabilities. The Macintosh graphical user interface (GUI), which utilizes pull-down menus and is relatively consistent from one software program to another, is credited with much of the Macintosh's success.

Commodore Amiga: The most video-friendly PC, inexpensive software, but poor marketing and limited user base. New hardware and software with the unusual name of Video Toaster™, by a company called NewTek, turns the Amiga into a remarkable graphics, switcher, and digital effects box and is turning the low-end production market on its ear. It will be interesting to see how these developments affect the professional broadcasting scene.

PCs in Audio Production

MIDI is a standard for computers to talk to synthesizers and other digital devices capable of creating audio signals. Sampling and **sequencing** software enable computers speaking MIDI to become effective tools for the creation of soundtracks for radio, television, and film productions. Another area of audio production ripe for computers is digital audio post-production. The *digital audio workstation (DAW)* became the buzzword for the late 1980s. These units integrate hard disk storage of digitized audio with computers that allow audio processing on a par with word processing. Audio segments can be edited, time compressed or expanded, or manipulated in any of a myriad of ways. At the high end of this technology is something called The Tapeless Studio®. Hard disk storage has made possible the elimination of multitrack audio tape recorders for audio production work. Instead, massive hard disks are wired together to allow gigabytes of storage. The user-definable sampling rate actually determines how many seconds or minutes of recording time are available given a fixed amount of digital storage. One such system is New England Digital's Synclavier® system, with its Direct-to-Disk® recording option. Personal computers prevalent in the audio production and post-production business include the Apple Macintosh, IBM and IBM-compatible PCs, the Commodore Amiga, and the Atari ST and Mega 4 PCs.

PCs and Computer Graphics

Personal computers are also used for computer graphics, both to try out new ideas and concepts and to create the finished product for broadcast. Paint and animation software for the Commodore Amiga, Apple Macintosh, and IBM compatibles now permit PCs to emulate everything from a simple character generator to a sophisticated paint box with full 3-D animation at 30 fps. For high-end professional applications, it may be necessary to process images through expensive outboard gear, but simpler, lower resolution output is now available directly from computers outfitted with the necessary cards and boards. Some industry analysts predict that in a few years the majority of computer graphics work will be produced on personal computers.

Although personal computers are being used more and more for video applications, most require additional outboard or internal video encoders to input and output NTSC video. For broadcast applications, be sure that the video encoder meets RS-170A specifications, or you may find that, although you may be able to achieve a stable recording, the video signal will not allow processing by video switchers, time base correctors, or other equipment necessary for legal broadcast. Another important concern for those using PCs for video work is the matter of genlock. If the output of a PC is to interface or be mixed with other video signals, the PC must be locked to a stable source of sync, usually the incoming video from a camera, time base corrected VTR, or switcher. Without genlock capabilities, it is impossible to use PC-generated graphics as a key source or fill with externally generated video signals.

Digital Compression

Digital data compression is a very important technology right now. Its importance for the broadcast industry is focused in two areas: transmission of digital signals and storage of digital signals. All data compression procedures rely on some type of encode/decode algorithms to eliminate redundant information. This, of course, has to be done carefully so that no important information is lost in the process.

Audio compression is critical to the success of digital audio broadcasting. At the rates required before compression, the transmission of a stereo audio signal is considerably spectrum intensive. Digital workstations also benefit from reductions in digital audio requirements as storage demands increase. As mentioned earlier, a mono audio signal sampled at 44.1 kHz requires 88.2 kB per second, or 29 MB per minute. To work with even relatively modest lengths of stereo audio, digital workstations require massive storage capability, usually on magnetic hard disks.

New compression algorithms that eliminate redundant information are changing storage and transmission requirements. Compression of the audio signal permits increased storage on smaller drives, or, looking at it another way, more minutes of music on your hard disk. Likewise, audio transmissions can now use lower capacity lines. Thanks to the phone companies, who have been working on compression for years in order to transmit more phone calls over limited lines, audio compression has come a long way. Stereo 16-bit PCM audio once required a transmission bandwidth of over 1.4 MHz; with new compression techniques, the same signal can now be carried over 64, 96, 128, or 256 kilobits per second (kb/s) lines, depending on the sampling frequency. MUSICAM, the compression scheme used by the Eureka 147 digital audio broadcasting system, requires 128 kb/s for its CD-quality audio.

Compression techniques are being used for digital video signals as well as audio. Like audio signal compression, video compression is of concern for digital storage as well as digital transmission. One place where digital storage of video images is critical is the area of off-line random access using magnetic disk technology. Systems such as the Avid Media Composer and the Emc2 compress video images before storing them on high-capacity computer hard disks. As discussed in earlier chapters, the video signal is a very complex one. A second of full-bandwidth NTSC video, less sync information, requires about 29 MB of storage. At that rate, the storage capacities currently available would quickly become full. As a result, video compression is used to increase capacity. One of the more recent advancements in video compression is the result of research by the Joint Photographic Experts Group (JPEG). Their CL550-30 image compression very large-scale integrated (VLSI) chip performs compression at ratios up to 100:1.

Transmission of a digital video signal is also an area of great concern for broadcasters. Whether they're talking bandwidth or transmission rate, broadcasters are interested in reducing the cost of transmitting a signal through air, wire, or fiber. A nine-bit, digital NTSC signal requires a 125-Mb/s transmission rate if no compression is used. Similarly, an NTSC component video signal requires 250 Mb/s. Looking ahead to future needs, a component HDTV signal will likely require a 1188-Mb/s data rate. However, due to the limitation of the human eye, compression systems that capitalize on pixel redundancy can reduce the transmission rate for NTSC video to 45 Mb/s with little apparent effect on picture quality. Still, compared to the 64 kb/s required by a telephone call, video requires a data rate nearly 700 times greater. However, the fiber optic cable that may someday deliver telephone, television, and data into the home will most likely employ even lower bit rates than those listed here due to improved compression techniques. Systems are currently being tested that promise bandwidth reduction by a factor of five or ten. One of the leading compression systems is General Instrument's DigiCipher video compression scheme. Video compression technology, though, has more far-reaching effects than just fiber optic transmissions of television signals. Satellite, cable, and broadcast transmissions of NTSC and even HDTV signals may soon be making use of this innovative technology to increase capacity and to provide an ever-increasing number of program services.

Self-Study

Questions

1. Which of the following is not a magnetic medium used to store digital information in a computer system?
   a. floppy disk
   b. hard disk
   c. CD-ROM

2. Which of the following is the time-discretionary part of the analog-to-digital conversion process?
   a. sampling
   b. quantizing
   c. aliasing

3. This computer peripheral performs modulation and demodulation of a digital signal data transfer:
   a. modem
   b. keyboard
   c. tape transfer

4. For analog-to-digital conversion of NTSC video, the sampling frequency normally used is four times the subcarrier frequency, or:
   a. 8.8 MHz
   b. 14.3 MHz
   c. 20 MHz

5. How many numerical combinations are possible with an eight-bit binary system?
   a. 8
   b. 64
   c. 256

6. Which of the following terms describes the result of low sampling frequencies during the A/D conversion process?
   a. error correction
   b. quantizing
   c. aliasing

7. Which of the following computer parts would not be considered an I/O device?
   a. CRT
   b. ROM
   c. mouse

8. The Nyquist rule suggests a sampling rate of:
   a. at least twice the frequency of the highest frequency you wish to sample
   b. at least 44 kHz
   c. 3.58 MHz

9. Which of the following digital audio recording formats is available to consumers with built-in copy protection technology?
   a. DASH
   b. DAT
   c. Pro-Digi

10. This digital videotape recording format uses 19mm-wide tape, records a component video signal, and is used for extremely high-end video production:
    a. D-1
    b. D-2
    c. D-3

11. Sampling and sequencing are terms often associated with the communications protocol MIDI. How is MIDI used in the broadcast industry?
    a. It is used to drive external devices in video post-production.
    b. It is used to drive electronic musical instruments for the purpose of composing and performing audio programs.
    c. It is used for station automation and robotics.

Answers

1. a. No. Floppy disks are a very common magnetic storage medium for computers.
   b. No. Computer hard disks, also known as Winchester disks, are a commonly used magnetic storage medium.
   c. Yes. Compact Disc Read-Only Memory (CD-ROM) is an optical device used to provide large quantities of data for computers. As its name implies, it is currently a read-only medium.

2. a. Yes. The sampling rate determines how often the analog signal is measured.
   b. No. Quantizing is considered to be the amplitude-discretionary process in A/D conversion.
   c. No. Aliasing describes the errors that result from insufficient sampling or quantizing rates.

3. a. Yes. The modem is frequently used to transfer digital data over analog phone lines.
   b. No. The keyboard is simply a common input device.
   c. No. Try again.

4. a. No.
   b. Yes. Subcarrier is 3.58 MHz, which, multiplied by 4, is equal to approximately 14.3 MHz.
   c. No. The correct answer can be found by multiplying the subcarrier frequency, 3.58 MHz, by 4.

5. a. No. The correct answer is found by using the formula 2^x, where x is the number of bits.
   b. No. The correct answer is found by using the formula 2^x, where x is the number of bits.
   c. Yes.

6. a. No. Error correction is actually a process that may be used to correct for some of the errors caused by low sampling frequencies.
   b. No. Quantizing has to do with assigning a value to the sampled analog waveform. Try again.
   c. Yes. Many broadcast computer devices include error correction algorithms or circuits and are marketed using the term anti-aliasing.

7. a. No. The CRT, or video monitor, is probably the most common output device connected to most computers.
   b. Yes. Read-only memory is considered to be neither an input nor an output device. Rather, it is memory, usually permanent, that is used to hold a specific instruction set used to operate the computer.
   c. No. The mouse is an input device commonly found on graphics oriented computers.

8. a. Yes. That is why audio signals, which have a range of 20 kHz, are usually sampled at a rate of more than 40 kHz.
   b. No. The Nyquist rule does not specify a sampling rate but rather a rate that holds a certain relationship to the frequency being sampled.
   c. No. This is the frequency of NTSC color video's subcarrier.

9. a. No. Digital audio stationary head is a professional recording format and does not have built-in copy protection.
   b. Yes. In fact, it was the inclusion of the copy protection device that finally opened the door to the marketing of DAT to consumers.
   c. No. Pro-Digi is a professional digital audiotape recording format and does not employ any copy protection device.

10. a. Yes. D-1, which is based on CCIR Recommendation 601, is presently considered to be the ultimate videotape recording format for high-end graphics and video post-production work.
    b. No. D-2, which uses the same 19mm tape as D-1 does, records a composite video signal rather than a component signal.
    c. No. D-3 utilizes 1/2-inch videotape and records a composite video signal. It is being marketed as a format to be used from acquisition through post-production and on-air.

11. a. No. The communication protocols most commonly used for the control of post-production peripherals are RS-232, RS-422, and SMPTE time code.
b. Yes. Musical instrument digital interface allows the user to program the automated playback of a variety of sound synthesizers. It is being used heavily in the music scoring and production business.
c. No. Try again.

Projects

Project 1

Survey your production facility and list the devices that use computer technology, interface with computers, or have built-in microprocessors.

Purpose

To learn the extent to which computer technology has been integrated into the broadcast industry.

Advice, Cautions, and Background

1. Many of the computers and microprocessor-controlled devices will be obvious, others will not. Do not overlook those that may have hidden computer technology. Examples are audio digital effects devices, video time base correctors, digital video effects boxes, edit controllers, and of course automation systems.

2. Take note of devices that incorporate the personal computer. What family of PCs is most commonly used? If you are in the music business, it may well be the Apple Macintosh. If you work in the television business, you'll likely find the IBM family or perhaps the Commodore Amiga. Is the user interface easy or difficult to learn?

How to Do the Project

1. List the devices found, and briefly describe the function provided by the computers/microprocessors.

2. Write a short paper explaining how more computer technology could be integrated into the present system to facilitate production and perhaps add a measure of automation.

3. Type up both parts of the project on a word processor. Turn the list and short essay in to your instructor for a grade.

Chapter 10
ADVANCED TELEVISION

Also known as *high-definition television (HDTV)*, *extended-definition television (EDTV)*, and *improved-definition television (IDTV)*, the idea of an improved television system seems to have as many faces as it has acronyms. Call it what you want, advanced television (ATV) is the hot television technology for the 1990s. No other technology today compares for excitement or controversy. ATV is heralded by some as the promising savior of America's electronics industry and by others as the U.S.'s biggest technological and economic failure of the century. Before going farther, it would be helpful to understand what ATV and HDTV actually mean.

First and foremost, ATV means a new visual experience for the viewer. With approximately twice as many scan lines as the current television systems, a larger screen with a wider aspect ratio, and CD-quality sound, the visual and audio experience will approach if not exceed projected 35mm film. According to CCIR Report 801, HDTV is described as being able to replicate reality when the viewer is seated three screen heights away from the display. Higher resolution, better color reproduction, separate color and luminance signals, a larger and perhaps wider screen, and life-like audio will all be combined to make the HDTV experience larger than life, especially when compared to the current NTSC system. Whichever system is adopted, it won't come cheap. The cost of consumer HDTV receivers, whether based on CRT, projection, or LCD technology, is likely at first to be priced beyond the range of the average consumer.

The major players in the ATV race are the global superpowers: Japan, Europe, and the United States. The Japanese, who claim to have been working on HDTV for over 25 years, have already begun HDTV service via their multiple sub-Nyquist sampling encoding (MUSE) system. Although the U.S. is primarily concerned with terrestrial broadcast of HDTV signals (due to concern for localism and protecting local broadcasters' interests), Japan and Europe are moving ahead with DBS delivery systems. Despite Japan's worldwide dominance in hardware and the U.S.'s role as the world's chief supplier of software, the nations of Europe are not about to roll over and accept whatever HDTV system Japan or the U.S. decides to adopt. In fact, Europe is already well on its way to developing its own HDTV (1250 scan lines; 50 fields per second) standard, which goes by the name *Eureka*. The European market has been at odds with the Japan Broadcasting Corporation (NHK) system for some time due to the incompatible frame rate (1125 scan lines; 60 fields per second). Europe is on a 50-hertz, 25-frame system with its PAL and SECAM systems. Converting from or to a 30-frame system is both costly and a technical compromise, according to European television hardware and software producers.

One-Step, Two-Step, Leap Frog . . .

In the United States, HDTV has garnered the attention of industry, government, military, education, and research institutions. Perhaps it is not surprising that the country with more than one-quarter of the world's total television receivers has been one of the slowest to come to grips with a workable plan for delivering this new technology. Americans, who have all but forfeited the consumer electronics industry to Japan, are trying to figure out a way to be a player in this new technology. In the economic, political, and business arenas, three scenarios for transition from NTSC to ATV are currently being debated: the one-step, two-step, and leapfrog approaches. One-step proponents argue for a quick and final decision on an HDTV delivery system so that the U.S. can get on with the business of making the transition. Two-step advocates believe that it is too early to make a final decision at this point. Instead, the U.S. should introduce limited and NTSC-compatible improvements now with the goal of achieving full HDTV technology a few years down the road, once things have had a chance to settle down. Leapfrog strategists argue for a stay of all current research and a focus on fully digital technology and possible delivery by fiber optics. This would allow the U.S to leap ahead of the Japanese and Europeans with a system that will take the nation into the next century. Proponents of the leapfrog approach include the computer industry, the telephone industry, and the Department of Defense (DOD). Of course, broadcasters are not about to sit by and watch other delivery systems bypass them. The stakes are high, and the players are not about to leave the bargaining table without a fight.

Perhaps the consumer electronics industry has the most to gain or lose. The value of television receivers currently in use is estimated at more than $120 billion. Whoever controls the technology for the high-definition system or systems that will replace the current NTSC, PAL, and SECAM systems will be in the best position to reap the monetary rewards. (Zenith is the only American company currently manufacturing television receivers in the U.S.) Not only is the electronics industry at stake. The microprocessor market, which is currently led by the U.S., is also at risk. Computer technology, which is crucial to some of the proposed systems, has largely been developed by U.S. corporations. The system that will eventually be endorsed or approved for the U.S. market is a subject for intense speculation and debate. However, the FCC has determined that no matter what HDTV system is adopted, it will have to be compatible with the current NTSC system.

The current HDTV debate really centers around three interrelated but separate areas of concern: production, distribution, and display.

Production

It should be understood at the outset that there is no reason why production and transmission must share the same system. In fact, as long as the production standard is readily convertible to the transmission standard, it makes a great deal of sense to use two different systems, according to many HDTV experts. For years, broadcast television has used 35mm film as its production format and as a source for transfer to NTSC video for transmission. By the way, 35mm film is really the only worldwide production or acquisition standard currently in use. It is ironic that all the talk of HDTV may have actually promoted the use of film as a production format. Due to all the uncertainty as to which HDTV system will finally prevail, many producers feel that the safest route is still to shoot on film; they reason that they will eventually be able to transfer the film images to whatever HDTV system wins out. Currently, the closest thing that the video community can promote as a worldwide production standard is D-1, which records both 525- and 625-line systems. In the HDTV arena, the quest for one worldwide production standard appears to have slipped from the grasp of those who are promoting the NHK standard, which is currently in use on several continents. The NHK system is being promoted as a worldwide standard with the assumption that once the material has been recorded and edited, it can be converted to NTSC, PAL, or even film. Once the HDTV distribution systems have been put in place, the 1125/60 video can be converted to whatever HDTV transmission system is required. The unknown variable here is the quality and cost of the conversion, neither of which can be adequately documented at this time.

1125/60

The current NHK standard of 1125/60 (1125 scan lines and 60 fields per second, 2:1 interlaced) was the first system used for both production and transmission. In the spring of 1991, NHK reported that more than 600 programs had been produced in the 1125/60 HDTV system. In what many consider to be a very controversial move, the SMPTE has approved the 1125/60 system as the HDTV production standard for the U.S., officially referring to it as *240M*. It is interesting to note that another standards-setting organization, the American National Standards Institute (ANSI), at first concurred with SMPTE but later withdrew its approval after an appeal by Capital Cities/ABC. SMPTE's stamp of approval may turn out to be a major boost to this system but keep in mind that SMPTE can and sometimes does approve multiple, competing standards. With an aspect ratio of 16:9, 1125/60 is definitely widescreen TV. However, it requires the use of 30 MHz of bandwidth for recording. To achieve these frequencies, the HDTV recorder incorporates a modified 1-inch machine that runs at twice normal speed. The 1125-line HDTV system is also in production in the United States. Rebo Productions, 1125 Productions, and David Nile's New York–based production company are just a few of those shooting in the 1125/60 HDTV format.

High-Definition Television Production in Progress

In the spring of 1988, CBS selected NHK's 1125/60 system for principal photography of the film *Innocent Victims*, the first U.S. made-for-TV movie produced using HDTV technology. By taping the program in HDTV instead of 35mm film, CBS estimates that it saved 15% of production costs. Made-for-television movies may appear to be a natural choice to begin the transition from 35mm film to HDTV for production, but what about films that will be released in theaters? The world's first high-definition theatrical film, *Julia and Julia*, was produced by the Italian network RAI. Another early HDTV production was the 14-part Canadian series, *Chasing Rainbows,* which was produced by the Canadian Broadcasting Corporation (CBC). The first theatrical feature film shot in the United States on HDTV, *Crack in the Mirror*, starred and was directed by Robby Benson. *Crack in the Mirror* was shot in HDTV by Rebo Productions and was transferred to 35mm film for release. All of these productions have been simultaneously praised and assailed on the basis of picture quality, production ease, and overall effectiveness.

Two glitches in the HDTV production process are still being resolved. One is constructing an imaging device that has both the resolution and the sensitivity necessary to produce an image suitable for HDTV pictures. Tubes are quickly being replaced by CCDs in almost every other production environment. However, tubes still have an edge in resolution, and resolution is, of course, central to the whole idea of HDTV. The tubes used in most HDTV cameras are high-gain, avalanche rushing-amorphous photoconductor (HARP) tubes. Unfortunately, resolution is achieved at the expense of sensitivity. The smaller the focus of the electron beam, the higher the resolution and the lower the sensitivity, thus requiring more light on the set. Especially when compared to the newer and faster 35mm film stocks, HDTV production requires extra lighting, which in production means more fixtures and increased setup time. The second complication has involved achieving the necessary optical resolution for the lenses used with HDTV cameras. HDTV lenses by Canon, Fujinon, and Nikon have been available for a short time but at a price. These lenses can cost more than $200,000!

NTSC—Room for Improvement?

Although it is an established fact that the NTSC video system has more than its share of problems, one thing that the HDTV debate has done has made television engineers look at the current NTSC system for improvements that could be made while retaining full compatibility. Two examples of attempts to improve the NTSC image include ATRC's Advanced Compatible Television and Faroudja Research's SuperNTSC™. Both systems tout their economical approach allowing for a gradual upgrade to full HDTV. These systems are sometimes referred to as *enhanced-definition television (EDTV)*. (*Improved-definition television [IDTV]* refers to those systems that concentrate their energies on the display of a standard NTSC picture and, through the process of line-doubling, achieve higher apparent resolution.) Using sophisticated encoding and decoding to alleviate many of the inherent problems with the standard NTSC signal, SuperNTSC is being positioned as an interim step in the conversion to advanced television. SuperNTSC was scheduled to be the first system to go before the FCC's advisory committee. However, in a recent change of events, Faroudja Research Enterprises has withdrawn SuperNTSC from testing. This decision may be due in part to the FCC's position, which has increasingly moved in favor of going to a full HDTV system rather than to an intermediate EDTV system. Another reason for SuperNTSC's withdrawal is that, according to Faroudja, SuperNTSC already complies with FCC regulation and does not need approval before being implemented.

Distribution

The problem of production is not nearly as complicated as that of distribution—in particular, terrestrial broadcasting. One of the principal problems is that of spectrum scarcity. As a general rule, the better an ATV system's performance, the greater its bandwidth needs. The NTSC video signal requires 6 MHz of spectrum space for terrestrial broadcast, and the full-bandwidth HDTV signal requires 30 MHz. (By using multiplexing and compression, researchers have been able to reduce this figure considerably.) Because the FCC is the regulatory body that controls the use of spectrum space, their approval is necessary before an HDTV transmission standard can be adopted for use

TABOOS: Broadcast frequency allocations which are restricted from use to prevent potential interference. The most common type of taboo is the co-channel taboo: Two different television or radio stations may not broadcast on the same frequency in the same geographical area. Another type of taboo ensures geographical spacing for adjacent channels, e.g., TV channels 2 and 3 may not operate in the same market.

in the U.S. The 1988 draft statement on HDTV by the FCC stated that any system to be considered must be compatible with the existing NTSC system. This means that the HDTV system to be adopted for U.S. broadcast must be able to transmit an NTSC-compatible signal within a bandwidth no wider than 6 MHz. One way to get around this requirement would be to use two channels. This additional spectrum would most likely have to come from existing unallocated "**taboo**" channels within the VHF and UHF spectrums. The controversy is compounded by the conflicting interests of several groups: notably, broadcasters, cablecasters, and, more recently, the phone companies. Fiber optics, satellites, and videocassettes are other technologies with the capability for the delivery of HDTV signals into the home. Terrestrial broadcasters are greatly concerned that HDTV pictures delivered by cable, satellite, or videocassette will make their pictures look bad by comparison. Analogies have been drawn comparing the transition from NTSC to HDTV to the rise of FM radio at the expense of AM. If that weren't enough, both broadcasters and cable operators are concerned that the introduction of fiber optics by the phone companies may allow the phone companies to begin delivery of additional services into the home.

The current NTSC-compatible HDTV systems encode the video in such a way that owners of standard NTSC receivers can receive an ordinary TV signal, and owners of HDTV receivers can decode the full HDTV picture. This is achieved in one of two ways: either the NTSC signal is augmented by a secondary signal that is broadcast on another frequency to complete the HDTV information, or the two signals are simulcast. In the first case, the HDTV receivers combine both signals to construct the HDTV image. In the second scenario, NTSC receivers continue to receive the NTSC broadcast while HDTV receivers tune into the simulcasted HDTV signal. In March 1990, the FCC indicated their clear preference for the simulcast rather than the augmentation approach. This has led to changes in some of the current proposals.

FCC Testing and Approval

The FCC Advisory Committee on Advanced Television Service has set up a testing schedule for the proponents of NTSC-compatible advanced television systems who would like to be considered for approval. Testing will take place at the Advanced Television Test Center (ATTC). Initially, testing was to have begun in the spring of 1990 and have continued through the fall of 1991. However, last-minute changes of several proposals to all-digital systems have resulted in the testing schedule being delayed. Testing is now scheduled to begin in July 1991. Despite the delays, it is expected that the FCC will choose a transmission system by mid-1993.

This is a game of high stakes, beginning with the testing procedure. Each of the system proponents had to pay a $175,000 testing fee just to reserve a position. The proponents are Advanced Television Research Consortium (ATRC is composed of the David Sarnoff Research Center, NBC, Thomson Consumer Electronics, and the North American Philips Corporation), Zenith Electronics Corporation and AT&T, Japan Broadcasting Corporation, and the American Television Alliance, which is made up of two formerly separate proponents: the Massachusetts Institute of Technology (MIT) and General Instruments Corporation. The four proponents are submitting six different systems for consideration by the FCC (ATRC and ATA are each submitting two proposals).

ATRC's Advanced Digital Television proposal is the only one of the systems to be tested that does not require an additional channel of spectrum space. The remainder are simulcast systems that require that the HDTV signal be co-broadcast on a 6-MHz channel adjacent to the 6-MHz NTSC signal. One of the simulcast systems proposed is NHK's Narrow-MUSE. According to its competitors, MUSE is subject to motion artifacts due to the subsampling technique employed. NHK's MUSE system is the only all-analog proposal at this time. Another proposal is Zenith-AT&T's Digital Spectrum-Compatible system. Scheduled for testing by the FCC in the fall of 1991, the Zenith proposal uses digital signal processing to fit 30 MHz of image into 6 MHz of bandwidth. By splitting the high- and low-frequency portions of the signal, the system is able to transmit much more signal with less power, which means less chance of interfering with the adjacent conventional channel. The ATRC's Advanced Compatible Television proposal will be the first system to be tested by the ATTC. Finally, there is the newly organized all-digital entry from American Television Alliance.

Perhaps the most striking development in the quest for an improved broadcast television standard has been the transition from analog to digital HDTV proposals. General Instruments' DigiCipher HDTV proposal generated interest in June 1990 when it became the first all-digital system. Since that time, other proponents have taken an all-digital approach. The Advanced Television Research Consortium announced its all-digital system in November 1990, and Zenith-AT&T switched over to an all-digital system in December 1990. This leaves NHK as the only proposal using analog transmission technology.

Display

One interesting thing about HDTV is that tests have shown that the average consumer does not notice much improvement over NTSC when viewing the images on a small display. For people to fully appreciate HDTV's advantages, a larger screen is required than the ones commonly used today. Futuristic scenarios depict wall-sized, flat-panel LCD displays of HDTV images, but unfortunately, these are still some years away. The choices today include direct-view CRTs and projection televisions. The largest direct-view monitors in production have just recently broken the 40-inch barrier. Sony has released a 43-inch direct-view NTSC monitor, which sells for $40,000. Even if you have that kind of money, there may be another hitch: most rooms and entryways are too small for a direct-view monitor of this size. A projection television may be easier to get into your house, but everyone has seen poorly aligned (and even properly aligned) projection TVs that displayed a poor image. Light output and resolution will have to increase before projection televisions are widely accepted.

Aspect Ratio

An attribute of HDTV closely related to screen size is aspect ratio. The aspect ratio of most HDTV systems is considerably wider than NTSC television. In fact, the wider aspect ratio is considered by manufacturers to be one of the most important attributes of

HDTV, especially when taking into account the mentality of the consumer. According to experts, for consumers to spend several thousand dollars on a new television receiver, there must be several visible and striking differences between the new technology and the old. The aspect ratio of the picture is one such difference. However, the perceived advantage of a widescreen experience comes with the disadvantage of receiver incompatibility. If the HDTV image is produced in a 16:9 aspect ratio, there would have to be some sort of aspect ratio accommodation to display the image on a 12:9 NTSC television receiver. The difference between 35mm theatrical film and NTSC television aspect ratios is solved in one of several ways. These including the letterbox approach, which preserves the integrity of the film's aspect ratio but introduces black areas at the top and bottom of the television frame, and the pan and scan approach, which compromises the film's aspect ratio but maintains a normal television image. Returning to the HDTV problem, those who propose the wider image must find a way to comply with the FCC's stipulation that for an HDTV system to be approved for terrestrial transmission, it must be able to deliver an NTSC-compatible image to households with the old receivers. The various HDTV proposals submit different methods to accomplish this. One proposal would increase the vertical blanking interval, somewhat akin to the letterbox approach, while another would implement a production technique known as *shoot and protect*. In either case, widescreen images would be delivered to the HDTV receivers while NTSC receivers would receive an image that encompasses the essential action.

Interlaced vs. Progressive

One unresolved matter for HDTV picture creation and display is whether to select interlaced rather than progressive, or sequential, scanning. Interlaced scanning, which is currently used by all worldwide television systems including NTSC, uses two fields to make up each frame of video. The effect is higher dynamic (motion) resolution while conserving precious bandwidth. Each field of video has only half the resolution of the entire frame, but because each field is replaced at twice the frame rate, the movement appears more fluid, and less flicker results. Sequential scanning systems do not divide the frame into two or more fields but rather increase the frame rate to reduce flicker and other motion artifacts. Sequential scanning more accurately approximates the motion picture imaging process. Although this increases the bandwidth requirements of the sequential scanning systems, the dynamic (moving) and temporal (static) resolution are improved over the interlaced systems. Working in partnership with AT&T Bell Laboratories and AT&T Microelectronics, Zenith submitted a proposal for a progressively scanned 787.5-line, 59.94-field system. Be careful when comparing horizontal resolution of sequential and progressive scanning systems. Progressive scanning of 787.5 lines will approximate a resolution of 1575 horizontal lines on a system which uses interlaced scanning.

Computer Monitors

William Schreiber, director of the Advanced Television Research Program at MIT, has proposed an innovative concept for ATV receiver design. At the center of his concept is a "smart" receiver, i.e., one with a computer for a brain. The computer technology would provide the capacity to adapt to a variety of television signals, and an open-architecture design would allow consumers to upgrade their system with each advance in the television transmission format. The open-architecture receiver would, in theory, protect consumers from obsolescence while avoiding the technical limitations of imposing one ATV system to which everyone must adhere.

Displays for the Department of Defense— Monitors for the Military

One group that is very interested in the high-resolution display of video signals is the U.S. Armed Forces. The Department of Defense (DOD) has commissioned its Defense Advanced Research Project Agency (DARPA) to investigate high-definition technology. Since 1989, DARPA has distributed $110 million to corporations invoved in high-definition systems research. DARPA's interest in HDTV goes beyond the obvious: it is also interested in maintaining the stability of the U.S. industries that produce electronics and computer chips essential to U.S. defense. DARPA hopes to use high-definition technology for computer displays and sophisticated simulations.

Self-Study

Questions

1. Under the larger category of advanced television, this improvement of the NTSC video image is designed to provide higher resolution now and perhaps act as a stepping-stone to greater improvements with an entirely new system in the future:
   a. EDTV
   b. HDTV
   c. MUSE

2. Unlike Japanese and European advanced television systems, the U.S. is looking to this type of signal distribution as its primary one:
   a. DBS
   b. cable
   c. terrestrial broadcast

3. The SMPTE-approved HDTV production standard has an aspect ratio of:
   a. 12:9
   b. 15:9
   c. 16:9

4. This EDTV scheme proposed by Faroudja labs is both channel-compatible and receiver-compatible, and it uses comb filtering and line-doubling techniques to improve the picture:
   a. 1250/50
   b. SuperNTSC
   c. MUSE

5. The FCC has made one major stipulation regarding its approval of an advanced television system. The FCC requires that the approved system:
   a. must favor U.S. television receiver manufacturers
   b. must be NTSC-compatible
   c. must have at least twice the resolution as NTSC

6. Currently, the most widely used HDTV production format is:
   a. NHK's 1125/60
   b. Eureka
   c. Advanced Compatible Television

7. Due to the uncertainty surrounding the acceptance of an HDTV production standard, many network and prime-time program producers continue to shoot on:
   a. NTSC video
   b. 35mm film
   c. SuperNTSC

8. The economics driving the ATV industry really revolve around the replacement of the nearly $120 billion worth of receivers currently in use. Which country currently stands to benefit the most from the sale of new receivers/monitors?
   a. U.S.
   b. Japan
   c. European Common Market

9. An ATV receiver proposed by William Schreiber of MIT would utilize a "smart" receiver that could receive any one of a number of ATV signals and would be user-upgradable. His proposal relies heavily on this technology.
   a. fiber optics
   b. LED
   c. computer

10. To maintain NTSC receiver compatibility, some ATV proponents propose using an augmentation channel, and others propose this:
    a. a simulcast system
    b. digital compression
    c. subcarrier encoding

Answers

1. a. Yes. Extended definition television employs techniques such as line-doubling to achieve higher resolution while retaining full compatibility with standard NTSC.
   b. No. High-definition television implies a new standard, different from and offering radical improvements over NTSC.
   c. No. *Multiple sub-Nyquist sampling encoding* is the name of several DBS transmission standards developed by NHK.

2. a. No. Although Japan and Europe depend heavily on DBS for the distribution of ATV signals, the U.S. is looking for a way to keep current broadcast stations in business.
   b. No. Although cable operators would like to be involved in the selection of an ATV production and distribution format, they are not being viewed as the primary suppliers of the new signals.
   c. Yes.

3. a. No. This is the aspect ratio of standard NTSC.
   b. No. Try again.
   c. Yes. At this aspect ratio, the appearance of HDTV images will more closely approximate theatrical projection.

4. a. No. This is the ATV standard proposed by the European community, which goes by the name Eureka.
   b. Yes.
   c. No. See **1c** for definition of *MUSE*.

5. a. No. The FCC has made no attempt to protect the interests of U.S. consumer hardware manufacturers.
   b. Yes. Although compatibility may be achieved by a variety of means, the FCC has insisted that the switch to an ATV system must not leave U.S. viewers without an alternative to buying a new television receiver.
   c. No. The FCC has made no specification regarding the performance of the proposed systems.

6. a. Yes. The 1125/60 production format is currently in use in a number of production companies around the country and in Japan.
   b. No. Eureka is the European community's ATV transmission scheme.
   c. No. Advanced Compatible Television has not yet gone beyond the prototype and testing stages.

7. a. No. NTSC video is not expected to translate well to ATV and therefore is not a viable standard for productions intended for future ATV distribution.
   b. Yes. Due to the high resolution of film and its acceptance as a worldwide standard, 35mm film is positioned as a safe choice for a production standard. Once the dust settles, film can be transferred to the accepted ATV standard.
   c. No. Try again.

8. a. No. Currently, there is only one remaining receiver manufacturer in the U.S.: Zenith.
   b. Yes. Most of the receiver manufacturers are located in Japan, and they have much to gain when (if) a standard is adopted and consumers begin to buy the new receivers.
   c. No. Although Philips, based in The Netherlands, is one of the largest television receiver manufacturers, most receivers are produced by Japanese companies.

9. a. No. Although fiber optics may well be the transmission medium of the future, it has nothing to do with the receiver's ability to adapt to dissimilar ATV signals.
   b. No. However, LEDs are being used in small video displays in place of electron beam scanning devices.
   c. Yes. It is the incorporation of computer technology that would allow the receiver to switch between various systems. Also, a simple software upgrade would allow incorporation of new developments.

10. a. Yes. The premise behind the simulcast system is that the NTSC receivers will continue to receive the NTSC transmission while the new receivers will tune into the separate HDTV transmission. At a later point in time, the NTSC transmission will eventually be discontinued.
    b. No. Try again.
    c. No. Try again.

APPENDIX

Technical Milestones in Radio and Television

1877

Thomas A. Edison invents the phonograph

1895

Guglielmo Marconi of Italy transmits and receives wireless signals; Alexander S. Popoff accomplishes same feat in Russia

1897

German scientist, Karl Ferdinand Braun, constructs first cathode ray tube scanning device

1907

Boris Rosing in Russia and A. A. Campbell-Swinton in England simultaneously develop methods of image reproduction using electromagnetic scanning

1912

Institute of Radio Engineers is formed

1915

Radio is demonstrated at the San Francisco World's Fair

1920

Broadcasting begins, first radio receivers advertised for sale

1922

First commercial broadcast, WEAF, New York. More than 500 stations begin broadcasts

1923

National Association of Broadcasters formed
First transatlantic broadcast
Dr. Vladimir K. Zworykin applies for patent on iconoscope, the first TV camera tube

1927

Philo T. Farnsworth applies for patent on electronic television system
Bell Telephone Laboratories demonstrates wireless television between Whippany, N.J., and New York

1928

First experimental TV station permits issued

1934

FCC established

1939

Black-and-white television introduced at New York World's Fair by RCA
First TV sets offered for sale
First experimental FM radio stations go on air

1941

Commercial FM radio operations begin, 21 stations licensed

1947

Transistor demonstrated by Bell Telephone Laboratories
First magnetic tape recorders sold in the U.S.

1948

45 rpm and LP records are introduced

1952

70-channel UHF band added to television broadcasting spectrum

1953

FCC adopts color television standards proposed by NTSC

1954

Color television broadcasting and receiver production begun

1956

Ampex introduces the first videotape recording device at the NAB convention;
    Quadruplex VTR records black-and-white television signal on 2-inch
    videotape
On November 30, CBS makes first use of videotape on network television with
    rebroadcast of evening news for West Coast

1958

Stereo records and phonographs introduced

1959

Audio cartridge recording system introduced at NAB convention by Collins Radio

1960

Toshiba proposes a videotape recording using the helical scanning process

1961

FM stereo transmission system approved by FCC

1962

Philips introduces audiocassette tape player
Telstar communications satellite provides first international relay of TV pictures
Congress passes law requiring all-channel tuning (VHF and UHF) in all television
    receivers

1963

TV transmitter remote control authorized by FCC
ITFS established by the FCC
Ampex introduces Editek system, which uses an audible cue tone to facilitate
    videotape editing

1964

RCA videotape cartridge developed
TEAC provides slow-motion color video playback system for NHK coverage of
    1964 Olympics

1965

> *Early Bird*, first international communications satellite, launched (Intelsat I)
> Sony introduces the portable BV-120 "Video-Corder," which weighs in at
>   145 pounds
> First robotic cartridge videotape recorder (RCA TCR-100) introduced
> First U.S. broadcast of freeze-frame and instant replay

1966

> First full-color TV network (NBC); NBC also is first network to broadcast
>   slow-motion video
> Dolby sells first noise reduction system

1967

> PAL/SECAM standards announced
> First "timecode-like" editing system for video, On-Time, is developed by CBS,
>   Hollywood, and EECO Corporation
> Solid-state imaging technology demonstrated
> Sony introduces first 1/2-inch portable VTR
> First electronically generated characters appear on broadcast TV
> First color slow-motion disk system (ABC and Ampex)
> Ampex introduces first color VTR
> NHK is first to digitally record audio signals

1968

> CBS uses a portable mini-cam for political convention coverage
> Trinitron tube developed
> One-inch Plumbicon developed
> FCC assumes jurisdiction over Community Antenna Television (CATV) systems
> First azimuth videotape recorder developed by Panasonic

1969

> SMPTE/EBU time code standard established to end the chaos of incompatible time
>   codes for various editing machines
> Neil Armstrong walks on the moon (July 20); worldwide audience watches the
>   event live
> Global satellite coverage is achieved

1970

> Color-under recording used in first 3/4-inch VTR from Matsushita, Victor
>   Company of Japan, and Sony
> Corning develops first low-loss optical fiber (20 dB/km)
> First floppy disk developed

1971

> U-format introduced by Sony and JVC
> CMX formed after a joint experiment between CBS and Memorex; releases first
>   random-access video editing system, the CMX-600
> NHK (Japan) begins experiments with high-line-number TV systems and discusses
>   the feasibility of a 1125 line system
> RCA joins EECO to develop and market the TCE 1000, an electronic editing
>   system based on time code
> First microprocessor invented

1972

Time base corrector introduced by Consolidated Video Systems, on sale at 1973
NAB convention for $8750

CMX 300, the first computerized editing system, is introduced for on-line editing
and auto-assembly of pre-edited shows

MCA gives first public demonstration of laser videodisk

First prerecorded videocassette tapes offered to consumers

Canada's Anik is the first geosynchronous communications satellite

MCA and Philips demonstrate first laser videodisk

1973

One-inch VTR (A format) shown by Ampex

First ENG cameras, Ikegami HL-33, used in electronic field production

Lexicon Varispeech introduces first audio time compression/expansion with pitch
correction

First fiber optic communications system installed

IBM introduces first Winchester drive

1974

First microprocessor used in broadcast equipment

Two-third-inch Plumbicon camera tube developed

Sony introduces Betamax home VCR and VO 3800, first ENG cassette recorder

CBS News begins electronic news gathering (ENG)

NBC purchases two NEC frame synchronizers (FS-10), the first product of its kind

First digital video effect (image compression and positioning)

1975

B format 1-inch VTR shown by Bosch-Fernseh

HBO begins program distribution via satellite

RCA demonstrates a CCD video camera, makes broadcast quality an objective

First personal computer introduced

1976

Sony shows 1-inch VTR (BVH-1000) at NAB convention in Chicago

Ampex shows VPR-1 helical recorder with automatic scan tracking; also
introduces a portable model, the VPR-10

Ampex shows first electronic still-store system, the ESS

VHS home recording format introduced

First North American low-power TV station, Pickle Lake, Ontario

1977

PBS begins operation by satellite

SMPTE type C format 1-inch VTRs introduced

TEAC introduces PCM digital audiodisk recorder

1978

NHK experiments with HDTV via satellite relay

NHK begins multiple-audio-channel television in Tokyo

Fiber-optic technology demonstrated

Digital VTR demonstrated

Cellular phone service begins in Chicago

1979

Portable B and C format VTRs with battery power shown

CCD telecine introduced by Bosch

National Radio Systems Committee formed to study AM stereo and FM technical
standards

Sony introduces Walkman personal audio device

1980

First of second-generation type C machines introduced

World standard for optical digital audio disk established

Closed-captioning of TV programs for hearing impaired begins

Ampex introduces first video paint system, Ampex Video Art (AVA)

Quantel introduces the first digital video rotation effect

1981

Sony introduces Betacam format VCR

M format introduced by Matsushita, RCA, and Ikegami

Half-inch Plumbicon and Saticon tubes introduced by Philips and NHK

HDTV demonstrated in U.S. at SMPTE convention in Los Angeles

TEAC develops optical laser write/read disk system

Ampex introduces ADO digital video processor

First camcorder ENG systems shown at NAB convention: RCA's Hawkeye,
Panasonic's Recam, and Sony's Betacam

Ikegami and Panavision introduce electronic cinematography cameras

Sony announces first high-definition video recorder

IBM introduces its personal computer

1982

FCC approves AM stereo broadcasting, low-power television, and DBS

Quantel's Mirage, a video effects device, introduced at NAB convention

NEC introduces DVE device at NAB convention

Bosch shows first 1/4-inch camcorder in prototype form (KBF-1)

CMX/Orrox shows a disk-based random-access editing system

Boston station WNEV commits to Panasonic's 1/2-inch M format for
production/post-production of its magazine show

Sony introduces Hi-Fi audio tracks recorded as FM signals onto videotape

First CD player sold

One hundred and twenty-two companies agree to standardization of 8mm video
format

The Bell system divestiture announced

1983

Network radio distribution by satellite (ABC, CBS, NBC, and RKO) using digital
format

Multiple-audio-channel television system selected by EIA for U.S.

Ku-band satellite transmission for broadcast tested by NAB

Use of FM subcarriers deregulated by FCC

NEC introduces the SPC-3 CCD camera

Last NAB convention exhibit with quadruplex VTRs

1984

FCC approves use of AM subcarriers for broadcast and nonbroadcast functions

Montage Picture Processor introduced by Montage Computer Systems

Lucasfilm/Convergence show the Editdroid disk-based editing system designed to
emulate film-style editing

First stereo broadcasts take place: "Tonight Show" and partial coverage of
Olympic Games

FCC authorizes multichannel TV sound

Apple introduces the Macintosh computer

Cellular telephone service spreads nationwide

1985

Panasonic introduces M-II format at NAB convention

SMPTE and CCIR approve D-1 digital component recording format for worldwide
program exchange

First digital audio tape recorders demonstrated

NBC becomes first network to switch over to all-satellite distribution; also, is sole
network using Ku-band

RCA's broadcast division folds

Sony's JumboTRON is world's largest TV screen at ten stories tall

1986

Sony introduces DVR-1000 digital videotape recorder based on
CCIR 601 (D-1) standard

Sony also introduces first 1-inch videotape recorder with PCM digital audio,
the BVH-2800

Ampex, Thomson, and Bosch sign manufacturing and marketing agreements with
Sony to produce Betacam products

NBC announces purchase of M-II products from Panasonic

SMPTE forms ad hoc group on high-definition studio systems to document
specifications for 1125/60 HDTV

Ampex submits its composite digital format (D-2) to SMPTE for standardization

Dolby introduces SR noise reduction system

1987

Superchannel DBS service begins in the U.K.

Abekas introduces the A-64 digital disk CCIR 601 recorder

NEC introduces SR-10 solid-state video recorder

Advanced Television Systems Committee announces plans to conduct over-the-air
tests of HDTV transmission formats

SMPTE working group on HDTV approves 1125/60 standards document

NAB forms HDTV technology center to study future of television

Super VHS (S-VHS) introduced

1988

Ampex and Sony introduce D-2 composite digital tape machines

Europe's Eureka 95 HDTV system demonstrated at IBC in Brighton, England

NBC proposes a 1050/59.94 HDTV system with the backing of ABC, Zenith, Thomson Consumer Electronics, North American Philips, and others

Philips Laboratories demonstrates HDTV system designed for satellite transmission (HDS-NA)

3M decides that 1988 will be the last year it markets videotape for the 2-inch quad format

Julia and Julia, first HDTV feature film, released

Hi-8 videotape recording format is demonstrated

1989

First fully compatible three-dimensional TV broadcasts

American National Standards Institute rejects the 1125/60 HDTV production standard accepted by SMPTE

Japan begins regularly scheduled HDTV broadcasts via DBS

First experimental broadcast of HDTV in the US

Panasonic shows prototype of digital camcorder

1990

JVC discontinues 3/4-inch U-matic format, leaving Sony as sole manufacturer

1991

FCC begins testing of HDTV systems

Recordable CD hardware makes its debut

GLOSSARY

A

ADO: Short for *Ampex Digital Optics*. This is probably the most popular digital video effects device in use in post-production editing suites. *ADO* is sometimes used synonymously with *digital video effects.*

ALIASING: Undesirable "beating" effects caused by sampling frequencies that are too low to faithfully reproduce image detail. Aliasing is evidenced in video as jaggies, moiré, or video noise, and in audio as unwanted frequencies that result when filtering is not used to eliminate frequencies above the range of the sampling frequency.

AM: Short for *Amplitude modulation*. A means of superimposing an audio signal on a carrier radio wave for transmission. AM varies the amplitude of the radio wave in accordance with the signal being broadcast.

AMPLIFIER: An electronic device that increases the strength of a signal, usually the output of a preamplifier. Most amplifiers are used to drive loudspeakers or other monitoring devices. A power amplifier usually has few controls, and a control amplifier often includes integrated preamp and mixing facilities.

ATTENUATION: To reduce the strength of a signal. A fader is a variable attenuator, and a loss pad provides a fixed amount of attenuation.

AZIMUTH RECORDING: A method by which video information is recorded onto magnetic tape without the use of guard bands between tracks of information. Head gaps positioned at different angles can read closely spaced tracks without interference from neighboring tracks. All of the new, high-quality video recording formats make use of azimuth recording to achieve excellent response from smaller tape surface recording areas.

B

BACK FOCUS: An adjustment to the video camera's lens to ensure sharp focus of the lens in relation to the face of the imaging device. The adjustment ring is typically found on the rear-most part of the lens. Back focus should be adjusted before attempting any registration procedures.

BALANCED: A three-conductor system for carrying audio signals that reduces hum and signal interference over long cable runs. Balanced connectors, e.g., XLR or Canon connectors, typically have three contacts.

BANDWIDTH: A range of frequencies within which a signal or transmission is contained. The greater the bandwidth of a transmission channel, the more information it can carry. In television, bandwidth is usually expressed in MHz.

BEAM: The focused stream of electrons that scans the face of the camera's pickup tube and the monitor's picture tube.

BINARY: Mathematical representation of a number to base 2, i.e., with only two states: 1 and 0; on and off; or high and low. Requires a greater number of digits than base 10, e.g., 254 = 1111110.

BIRD: Another name for a satellite.

BIT: One digit of binary information. A contraction of the term *binary digit*, it has a value of either 0 or 1.

BLACK (COLOR BLACK or BLACK BURST): A composite color video signal made up of composite sync, burst, and a black video signal, normally at 7.5 IRE units above blanking.

BLACKED TAPE: A tape that has been prepared for insert editing by having stable video and control track recorded for its duration, or at least a bit longer than the length of the program to be edited. The term *blacked* is used because a video black signal is commonly used as the source of video and control track, although any stable video signal will do the job. The term *striped* is often used to refer to a blacked tape that has also been recorded with time code. It is important that the video, control track, and time code signals be recorded so that they are locked and synchronous with each other.

BLANKING: The portion of the video signal that turns off or *blanks* the scanning beam during retrace. Blanking has both vertical and horizontal components. On a waveform monitor, horizontal blanking is the signal information between active lines of video.

BLUE-ONLY DISPLAY: A feature that allows the CRT to display the output of only the blue gun. By defeating the red and green signals, accurate settings of hue and saturation can be obtained using a SMPTE color bar test signal. Using the blue-only feature, the operator need not have perfect color vision, simply the ability to discern the differences in brightness values of the bars containing blue information.

BURN (BURN-IN): A flaw in a camera pickup tube that results from extended overexposure. CCDs, unlike tubes, are immune from burn-in.

BURST: See COLOR BURST.

BYTE: One byte of digital video information is a packet of bits, usually but not always eight. In the digital video domain, a byte is used to represent the luminance or chrominance level. One thousand bytes is one kilobyte (kB) and one million bytes is one megabyte (MB).

C

CARDIOID: A unidirectional pickup pattern for microphones. Named for its heart-shaped pattern, which is most sensitive to sounds directly in front, less sensitive to sounds coming from the sides, and least sensitive to sounds from the rear.

CARRIER WAVE: The radio wave on which an audio or video signal is superimposed for broadcasting.

CAV: *Component analog video.* This term is usually used to refer to the video recording formats that use component processing and recording of an analog video signal on 1/2-inch magnetic tape. CAV formats include Betacam, Betacam SP, and M-II.

C-BAND: Typically operating at a frequency of 4 to 6 GHz, these low-power transponders operate on 5 to 10 watts of power (requiring receiving dishes about 3 meters across), although the next generation of C-band birds will have 16-watt transponders. Typically, each C-band satellite carries 24 to 36 transponders.

CCD: Short for *Charge-coupled device.* These solid-state imaging transducers are used in video cameras in place of pickup tubes. They provide numerous advantages over tubes and are expected to replace tubes entirely within a few years.

CCIR RECOMMENDATION 601: Encoding Parameters of Digital Television for Studios. The international standard for digitizing component color television video in both 525- and 625-line systems, for which it defines 4:2:2 sampling at 13.5 MHz with 720 luminance samples per active line and eight-bit digitizing.

CCU: Short for *Camera control unit.* This is the part of the camera chain that houses many of the control and setup functions of the camera. Usually located apart from the camera head (in the control room or remote truck), the CCU allows the cameras to be shaded during a real time multicamera production.

CENTERING: The rudimentary registration adjustment. Centering is the process of adjusting the H and V position of the red and blue tubes to ensure that the output of all three tubes aligns perfectly.

CHIP CHART: Also known as a *grey scale* or *logarithmic reflectance chart,* this camera chart typically has two grey scales, one ascending and the other descending. The chip chart is used when performing a manual color balance on a color camera.

CHROMINANCE: The color information part of the video signal, usually defined in terms of hue and saturation. The video signal is made up of chrominance (color) and luminance (brightness) information.

CLEAR CHANNEL: An FCC designation for a Class 1 AM station, which is permitted a maximum power of 50 kilowatts. The U.S. has 45 clear channel allocations.

COAX: Short for *coaxial.* A cable composed of a single conductor core surrounded by a braided shield. Coax is the type of cable typically used to route video signals from one component to another. RG-59 is a common grade of coax video cable.

CODING: The third step in the analog-to-digital process in which the information is written in binary form.

COERCIVITY: A measure of the magnetic field strength required to return the material to a state of zero magnetization, measured in oersteds (Oe). A typical range for videotape is 300 to 1500 Oe.

COLOR BALANCE: An adjustment of the output of the red and blue tubes or chips of a color camera to ensure that the camera faithfully reproduces white and all other colors. Color balancing includes white balance and black balance procedures and compensates for various color temperatures of light.

COLOR BARS: A video test signal used as a reference to set up and test video components. Color bars are available in several configurations, including full-field, split-field, and SMPTE standard. The standard array of colors is white, yellow, cyan, green, magenta, red, blue, and black.

COLOR BURST: Composed of nine cycles of subcarrier, this is a reference signal for interpreting the encoded color information. The color burst signal is visible in the horizontal blanking interval on the waveform monitor or on the 180-degree line on the vectorscope.

COMB FILTER: An electronic filter used to lessen the effects of cross-color, cross-luminance interference in the video signal.

COMPANDING: A noise reduction technique that uses a two-stage process: *compressing* the signal before sending (reducing its dynamic range) and then e*xpanding* the signal once it has been received to restore it to its original dynamic range.

COMPONENT: Y, R-Y, B-Y, or Y,U,V. A video signal in which luminance and chrominance information is kept separate rather than being combined as in the composite video signal. Component processing and routing requires three wires to route the signal, and component recording requires the use of separate tracks on the magnetic tape. Betacam and M-II are two popular component video recording formats.

COMPOSITE: Standard video that combines chrominance and luminance information by encoding the output of the red, green, and blue channels into the Y, I, and Q signals. Composite video also includes blanking and sync and is the standard for broadcast transmissions of video signals.

COMPRESSION: Reduction of dynamic range. Used in broadcasting to achieve greater or more uniform loudness. Digital compression involves the use of algorithms to reduce the bandwidth necessary to store or transmit a digital signal.

CONFIDENCE HEADS: Playback heads that allow monitoring of the recorded signal while the VTR is in the record mode. Confidence heads permit instant verification without the need to interrupt the recording in progress.

CPU: Central processing unit. This is the collective name for the primary memory, logic units, and the main control unit of a computer system.

D

dB: See DECIBEL.

DBS: Direct broadcast satellite. Satellites powerful enough (approximately 120 watts on the Ku-band) to transmit a signal directly to a medium to small receiving dish (antenna). DBS does not require reception and distribution by an intermediate broadcasting facility, but rather transmits directly to the end user.

DECIBEL: A relative unit of measure used to compare sound levels of one signal to another. One-tenth of a Bel (in honor of Alexander Graham Bell) is a deci-Bel or dB. One dB is the smallest increase in volume that can be perceived by the human ear. An increase of 6 dB equals twice the sound pressure and is perceived as a doubling of volume.

DEGAUSS: To apply a random magnetic field for the purpose of erasing magnetically stored information. A videotape or a CRT can be degaussed, in the first case to erase the tape and in the second case to remove magnetic charges that cause portions of the screen to be discolored.

DELTA GUN: A CRT design in which the three guns are positioned in a triangular fashion rather than in a line.

DIGITAL: An electronic system that processes information as a series of binary codes rather than an analog waveform.

DIN: Deutsche Industrie Normen. The German industry standard for electrical and electronic devices and connectors. European equipment manufacturers often use DIN connectors.

DIRECT WAVES: Radio frequency (RF) transmissions that are line of sight. Transmission distance is limited by curvature of the earth and by physical obstructions.

DISTORTION: Modification of an electronic signal that may or may not be desirable. In audio, overload distortion and total harmonic distortion (THD) are two common types of problem distortion.

DOLBY: British inventor, whose most notable invention was the Dolby noise reduction system. Dolby A, B, and C, and Dolby SR are all types of electronic processing used to increase signal-to-noise ratio and reduce unwanted frequencies, specifically the tape hiss inherent in the recording process.

DOWNLINK: A receiving dish. This could be a passive receiving antenna for a single household, in the case of DBS, or the antenna for the head-end of a cable system.

DROPOUT COMPENSATOR: A feature found on many time base correctors that inserts video information where a tape dropout occurs. The inserted video is taken from the preceding line, and although it does not match the missing material exactly, it is better than seeing a black or white streak.

DYNAMIC: A term used to describe signals that are undergoing changes in amplitude or frequency. Dynamic range has to do with the range between the extremes of amplitude or frequency of an audio signal. A dynamic microphone has a specific type of transducing element.

DYNAMIC TRACKING: Also known as *automatic scan tracking*. This feature permits stable playback at other than standard speeds. Most professional VTRs that have the dynamic tracking feature can play back in variable speeds ranging from −1 to +3 times normal speed.

E

EDL: Short for *Edit decision list*. This is a list of numbers signifying in and out points for each of the series of shots making up the program. The EDL may be a written list of numbers, a computer printout of time code numbers, or a computer floppy disk containing the same list of numbers in a format readable by the on-line editing computer.

E-E: Short for *Electronics to electronics*. This feature allows the VTR's record circuitry to be activated without engaging the tape transport. The VTR passes through all of the signals that are connected to its inputs. This allows the operator to monitor the video and audio levels downstream of the VTR.

ELECTROMAGNETIC SPECTRUM: The entire spectrum of frequencies of electromagnetic waves. The electromagnetic spectrum ranges from lowest to highest in the following order: radio, infrared, visible light, ultraviolet, X-ray, gamma ray, and cosmic ray waves.

ENCODE/DECODE: The process of converting video from its RGB components into composite video, and vice versa. Great improvements have been made in the area of encoding and decoding by Faroudja Laboratories, the result being improvement in the quality of encoded NTSC video.

EQUALIZER: A device that allows you to increase or attenuate certain frequencies of an audio signal. Graphic, selectively variable, and parametric are all types of equalizers. Often expressed as *EQ*.

EXPANSION: The increase of the dynamic range of an audio signal. The opposite of compression. (See also COMPANDING.)

EXTERNAL SYNC: This feature allows the monitor/scope to receive synchronization pulses from an external source (usually from the house sync generator) instead of taking sync from the incoming video signal. The use of external sync facilitates making timing adjustments.

F

FIBER OPTICS: A technology based on the conversion of electronic signals to digitized pulses of light that can be transmitted through bundled glass fibers. Some predict that fiber optics will someday replace coax, microwave, and satellite transmission technology.

FIELD: Half of a video frame, 262.5 horizontal lines (NTSC).

FLUTTER: An instability or variation of a mechanical transport, usually in a turntable or tape machine, and usually between 5 and 15 Hz. Introduces undesirable distortion to the signal.

FLUX: The generated magnetic field or lines of force that are produced by the recording head to magnetize the particles in the magnetic tape.

FM: Short for *Frequency modulation*. A means of superimposing an audio signal on a carrier radio wave for transmission. FM varies the frequency of the radio wave in accordance with the signal being broadcast.

FOOTPRINT: The geographic area covered by a satellite's transmission signal. Most of the surface of the globe can be covered by the footprints of three satellites strategically placed.

4:2:2: One of the ratios of sampling frequencies used to digitize the luminance and color difference components (Y, Cb, Cr) of component video. The luminance signal is sampled at 13.5 MHz (4 times 3.37 MHz), and the two color-difference signals (Cb, Cr) are sampled at 6.75 MHz (2 times 3.37 MHz).

4:4:4: One of the ratios of sampling frequencies used to digitize the luminance and color-difference components (Y, Cb, Cr) or the RGB components of component video. It differs from 4:2:2 in that all components are sampled at 13.5 MHz.

4:4:4:4: The same as 4:4:4, except that the key signal is included as a fourth component, also sampled at 13.5 MHz.

FRAME: One complete video picture. In NTSC, takes place in one-thirtieth of a second and is made up of 525 lines and two fields.

FRAME SYNC: Short for *frame store synchronizer*. This is a digital storage device capable of delaying an incoming, nonsynchronous video signal by as much as one frame to synchronize the signal with house sync.

FREQUENCY: The number of oscillations or cycles per second. When dealing with acoustic sound waves, frequency is expressed in hertz (Hz) and perceived as pitch. The frequency of a radio wave affects its physical behavior. One kilohertz is 1000 Hz; 1 megahertz is one million Hz; and 1 gigahertz is one billion Hz.

FREQUENCY RESPONSE: The manner in which an audio component responds to its source, i.e., whether certain frequencies are enhanced or attenuated. A microphone with a flat response responds equally to frequencies across a broad spectrum. A frequency response curve is a visual graph representing the performance of a piece of electronic equipment, e.g., a microphone.

G

GAMMA: Having to do with the grey tones of a video or film image. Some video cameras and their CCUs allow gamma to be adjusted, thus affecting the way that the camera reproduces midgreys.

GENERATION: A copy or replication of an audio or video signal. Each generation or copy is further removed from the original signal. In the analog realm, each generation introduces degradation of the signal quality; in theory, digital recording permits each generation to be exactly the same as its source.

GENLOCK: The locking of a video component or system to the synchronization pulses of an incoming signal. For example, a camera can be genlocked to a switcher or to another camera so that the signals from each can be mixed without any disruption.

GEOSYNCHRONOUS ORBIT: Satellites are placed in orbit 22,300 miles over the equator and remain in a fixed position above a specific location on the globe. The satellite remains in this fixed orbit, with a few minor adjustments from time to time, until its lifetime is expired (currently about 10 years, but soon to be increased to 12 years with the new generation of birds).

GPI: Short for *General purpose interface*. This is a computer editing protocol used to trigger other devices in the edit suite, such as the ADO, the switcher, or an ATR. The GPI is a parallel interface that can be set to trigger a device at a particular moment during an edit. The exact moment is usually defined by time code number.

GROUND WAVES: RF transmissions of a low frequency that travel through solid objects, i.e., ground or water.

H

HARD DISK: A device for magnetic storage of digital information. Also known as *fixed disks* or *Winchester disks*, hard disks record information on a spinning disk of smooth metal, usually aluminum. Hard disks have faster access time and contain a greater amount of storage than floppy disks.

HDTV: Short for *High Definition Television*. A technology that provides increased resolution, higher-quality audio, a wider aspect ratio and other improvements over the current television standards. The FCC is expected to make a recommendation for approval of an HDTV transmission standard sometime in 1993.

HELICAL SCAN: Also called *slant track*, this videotape recording method writes the information to the tape using diagonal or slanted tracks. The name comes from the word *helix*, which describes the fashion in which the tape wraps around the head drum.

HETERODYNE: This method of recording a video signal reduces the chrominance information to a lower frequency and narrower bandwidth before recording. The U-Matic format reduces the chroma from 3.58 MHz to 688 kHz, and VHS reduces it to 629 kHz. Also known as *low band* or *color under*.

HI-FI: Also known as *FM* or *AFM* audio, Hi-Fi audio is a feature available on certain consumer and professional video recording formats that provides extremely high fidelity audio recording specifications. One drawback, however, is the fact that the Hi-Fi audio is depth-recorded underneath the video signal and cannot be edited separately from the video.

HIGHBAND (DIRECT COLOR): A means of recording a video signal whereby the chrominance information is recorded at full bandwidth.

I

I/O DEVICES: Short for *Input/output devices*. These are the means for getting information into and out of a computer. The keyboard, mouse, monitor, and printer are all I/O devices.

IEEE: Short for *Institute of Electrical and Electronic Engineers*, formerly the Institute of Radio Engineers. The scale on a waveform monitor is divided into IRE, or IEEE, units.

IMPEDANCE: A characteristic of electrical components, rated in ohms and usually expressed as either Hi-Z (10,000 ohms and above) or Low-Z (50 to 300 ohms).

IN-LINE GUN: A CRT design in which the three guns are positioned side by side in a line, rather than arranged in a delta configuration.

IRE: Short for *Institute of Radio Engineers*. IRE units are used to measure the amplitude of a video signal. On a waveform monitor, blanking is at 0 IRE, white is at 100 IRE, and the tip of sync is at –40 IRE.

ITFS: Short for *Instructional Television Fixed Services*. This broadcasting service reserved for educational use is located in the 2500- to 2690-MHz range. Twenty-eight channels are available for ITFS to share broadcast programming within and among school systems.

K

KU-BAND: Operating in the 11- to 14-GHz range, these are medium-power satellites, requiring about 40 to 80 watts per transponder and permitting receiving dishes as small as or smaller than 1 meter across. Formerly, the demand for higher power limited the number of transponders to about ten per bird; however, the newer Ku birds have as many as 24 transponders.

L

LAN: Short for *Local area network*. A data communications system consisting of two or more microcomputers physically connected together with some type of wire or cable, over which data is transmitted.

LED: Short for *Light-emitting diode*. A type of semiconductor that lights up when voltage is applied.

LIMITER: An electrical device that prevents an audio signal from superseding a predetermined limit. Used to prevent overmodulation.

LINEAR: Having to do with recording formats or editing processes in which the material is accessed by shuttling forward or backward. Magnetic tape is a linear medium in that it requires that the operator search in a sequential fashion.

LONGITUDINAL TIME CODE: Time code that is recorded on the videotape on a linear audio track rather than in the vertical blanking interval of the video signal. Longitudinal time code is an audio signal that uses 80 bits to assign a numerical value expressed in hours, minutes, seconds, and frames to each and every frame of video.

LOW BAND (COLOR UNDER): See HETERODYNE.

LPTV: Short for *Low-power television*. These television stations were given permission to broadcast at low power on unused VHF and UHF channels. The VHF stations are limited to 10 watts of power; the UHF, to 1000 watts. This limits their effective range to 15 to 25 miles. The initial idea was to provide minorities and other special interest groups access to the broadcast media.

LUMINANCE: The brightness information part of the video signal. Luminance is often designated by the symbol Y. On a waveform monitor, the luminance level of the video signal can easily be measured by viewing in the L-Pass or IRE mode.

M

MICROPHONE: A transducer that converts sound energy into electrical energy.

MICROWAVES: That part of the electromagnetic spectrum lying between 300 and 300,000 MHz. Typically used by broadcasters for point-to-point, line-of-sight transmissions.

MIDI: Short for *Musical instrument digital interface*. This is the standard protocol for computers to interface with musical instruments (synthesizers) to permit sequential playback.

MIXER: An electronic component that allows several audio signals to be combined. Usually contains preamps, attenuators, and tone controls for each channel so that input signals can be controlled.

MODEM: Short for *Modulator/demodulator*. This device converts a signal from a computer to one capable of being sent over telephone lines. A modem at the other end converts the signal back to one able to be received by another computer.

MODULATION: The process of superimposing a signal onto a radio frequency carrier wave for transmission.

MONITOR: Audio: A transducer for converting electrical energy from an amplifier into sound energy. Video: A transducer for converting electrical energy into visible light.

MULTIPLEXING: A means of transmitting two or more signals over a single wire or carrier wave.

N

NTSC: Short for *National Television Systems Committee*. The committee formed to determine the guidelines and technical standards for monochrome and, later, color television. Also used to describe the 525-line, 59.94-Hz color television signal used in North America and several other parts of the world.

NYQUIST: Named for Harry Nyquist, the Nyquist rule states that to be able to reconstruct a sampled signal without aliases, the sampling must occur at a rate of more than twice the highest desired frequency. For example, CDs have a sampling rate of 44.1 kHz and allow signals up to 20 kHz to be recorded.

O

ODD-EVEN INTERLACE SCANNING: The method by which imaging devices and picture tubes create and display 262.5 lines of information 60 times each second. An electron beam scans all of the odd-numbered lines first (one field) and then scans all of the even-numbered lines (the second field), thereby creating an entire frame (two fields) composed of 525 scanning lines. It repeats this process 30 times every second to produce a flickerless image.

OFF-LINE: An edit suite or session in which the resulting product is an EDL, which in turn is used to expedite the on-line editing of the master videotape. Off-line edit suites are used to make creative decisions and to create a rough cut rather than the finished product.

OMNIDIRECTIONAL: A microphone pickup pattern that is equally sensitive to sounds coming from all directions.

ON-LINE: An edit suite or session in which the resulting product is the finished edited master videotape. On-line edit suites typically feature high-end VTR formats, allow sophisticated effects, and demand a higher rate than off-line suites.

P

PAPER EDIT: This phrase is used to describe the off-line editing process in which the producer keeps a hand-written log of the edits and their in and out points. This can be performed using a simple cuts-only editing system and window dubs of the time-coded source footage.

PCM: Short for *Pulse code modulation*. A means of digitally coding an audio signal for recording on magnetic tape. Videotape formats that use PCM audio recording include 8mm, Hi-8, and specific models of 1-inch type C.

PEDESTAL: Also known as *setup* or *black level*, this is the part of the video signal with the lowest level or IRE value. Pedestal is normally set at 7.5 IRE units above blanking (0 IRE on the waveform monitor) and appears as black on the video monitor.

PHASE: Typically, having to do with the timing relationship of two electrical or electronic signals. In audio, the phase or polarity of two or more signals that are being combined must be the same or low frequency response will suffer. In video, color phase corresponds to hue.

PICKUP TUBES: The former standard (see also CCD) for imaging devices in video cameras. A pickup tube is a transducer that converts light in electronic energy by means of an electron beam that scans a photosensitive faceplate. The most common tubes are the Saticon™ and Plumbicon™ tubes.

PIXEL: Short for *Picture element*. The smallest unit of measurement when dividing an electronic image. Commonly used to define resolution of imaging or display devices.

PREAMPLIFIER: An electronic device that boosts the output of low-voltage transducers (e.g., microphones and turntable styluses) before connection to a power amplifier.

PREROLL: This is the time it takes for a VTR to get up to speed in preparation for an edit. Most professional VTRs can get up to speed in as little as 3 seconds; however, 5 seconds is the normal preset duration for most editing systems.

PRISM BLOCK: This piece of optical glass acts as a beam splitter in a three-tube color video camera, separating the light gathered by the lens into its red, green, and blue components. The prism block has virtually replaced the dichroic mirror method of beam splitting.

PROC AMP: Short for *Processing amplifier*. The device that allows adjustment of the parameters of the video signal, e.g., video level, black level, chroma level, and chroma phase.

PULSE DELAY (PULSE CROSS): A feature on some monitors that delays the horizontal and vertical scan so that the H and V blanking intervals are displayed on the screen. Sometimes labeled as *H* and *V Delay*. Pulse cross permits the observation of the horizontal and vertical blanking intervals to check pulse widths and system timing. This feature is handy in lieu of a waveform monitor to inspect the blanking intervals.

Q

QUANTIZING: The process of sampling an analog signal to determine digital values equivalent to the voltage levels of the original analog signal.

R

RAM: Short for *Random-access memory*. A type of primary storage in which any randomly selected segment can be accessed at will for reading or writing.

RANDOM ACCESS: Having to do with formats or processes in which the material is accessible randomly and in the same amount of time. Random access of audio or video material is achieved by any of the disk-based formats, e.g., phonograph, CD, laserdisc, or computer hard disk.

RASTER: The portion of the camera pickup tube or CRT that is traced by the scanning electron beam.

REGISTRATION: The process of accurately aligning the output of the red, green, and blue imaging devices to achieve optimum resolution and color clarity of the resulting combined image. Registration involves several adjustment procedures for the red and blue tubes, the most common of which is known as *centering*.

RESOLUTION: Generally defined as the ability to convey detail or the number of pairs of black/white lines that can be distinguished. Camera imaging devices and CRTs exhibit varying degrees of image sharpness. Likewise, video recording devices may record and play back with similarly varying results, which are largely dependent on the frequency of the recorded signal. Engineers typically use a ballpark figure of 80 lines of horizontal resolution for each MHz of bandwidth. Using this equation, the 4.2-MHz bandwidth of a broadcast video signal allows approximately 330 lines of resolution. A camera's resolution is usually defined by its horizontal resolution, i.e., the number of vertical black/white line pairs that it is capable of resolving. An EIA RETMA resolution chart is commonly used to measure a camera's resolution capability.

RETENTIVITY: A measure of the flux density remaining after the external magnetic force has been removed.

RGB: Red, green, blue. Video in its purest form, as divided into its red, green, and blue components. Some computer graphics devices and monitors work with an RGB signal, but it is not a common format for recording or transmission technology.

ROM: Short for *Read-only memory*. A type of primary storage written to only once, after which it can be read from but not altered.

RS-232: Short for *Recommended Standard 232*. A communications protocol popular for serial communication between computer devices.

RS-422: Short for *Recommended Standard 422*. A communications protocol popular in the video post-production environment. Computerized editors and VTRs commonly use RS-422 interfaces.

S

SAMPLING: The process associated with analog-to-digital conversion in which the continuous analog waveform is divided into discrete moments in time. Also, recording a short segment of audio into digital memory to play it back as a musical element.

SENSITIVITY: The measurement of a video camera's ability to produce an acceptable picture in low-light conditions. Increased sensitivity is usually achieved by improvements to the lens quality, imaging device construction, and preamplifier circuitry. The sensitivity of a video camera can be artificially increased by engaging the gain switch. In effect this increases the output of the preamps, amplifying the video signal and any electronic noise.

SEQUENCING: Using a computer to order the playback of a series of electronic instruments. Sequencing allows a composer to build a musical track with multiple instruments playing together at one time under the control of a computer using MIDI software programs.

SC/H PHASE: Short for Relationship of *subcarrier to horizontal phase*. The industry has standardized SC/H phase so that the subcarrier phase is observed at the leading edge of horizontal sync. On line 10 of color field 1, it should be at 0 crossing and going positive. Some waveform monitors have LED displays to indicate proper SC/H phase.

SHADOW MASK: The metal grill between the guns and the phosphor screen in a CRT. The shadow mask prevents the output from each gun from activating the phosphors of another color.

SKIP WAVES: RF transmissions, most commonly shortwave and AM broadcasts, that bounce off of the ionosphere and can be received over great distances. Skip waves occur more frequently during nighttime hours.

SMATV: Short for *Satellite master antenna television*. A satellite broadcast service available to hotels, motels, apartment complexes, condominiums, etc.

SMPTE: Short for *Society of Motion Picture and Television Engineers*. This organization is actively involved in setting standards for the film and television industries.

SNG: Short for *Satellite news gathering*. A recent technology that supersedes ENG. SNG makes it possible to send live feeds from virtually anywhere in the world back to a local or network news bureau.

STICTION: The adhesion of two smooth surfaces (in this case, the magnetic tape and the head drum) in the presence of moisture. Most professional VTRs have a dew sensor that indicates excessive moisture in the air (humidity) or on the head drum. The VTR's circuitry prohibits the VTR from being operated until the conditions improve.

SUBCARRIER: A continuous sine wave used to encode the color information into the video signal. Subcarrier has a frequency of 3.58 MHz in NTSC video.

SYNC: Short for *synchronization pulses.* Horizontal and vertical sync pulses drive the scanning process at the camera and picture monitors. Composite sync (H and V sync combined) is sometimes required by video components to interface properly with other components. A sync generator is a master source of sync for a facility.

T

TABOOS: Broadcast frequency allocations which are restricted from use to prevent potential interference. The most common type of taboo is the co-channel taboo: Two different television or radio stations may not broadcast on the same frequency in the same geographical area. Another type of taboo ensures geographical spacing for adjacent channels, e.g., TV channels 2 and 3 may not operate in the same market.

TALLY: A light, usually red and mounted on either a camera head or monitor, that indicates the "on-air" source.

TELCO: Slang for the land lines leased from the telephone companies. Before the breakup of the Bell system, TELCOS were a much more common part of a local station's or a network's transmission process.

TELECONFERENCING: Also known as *videoconferencing,* this is the use of audio and still or live video images as a means of conducting meetings between distant locations. The current technology uses microwave, satellite, TELCO, or fiber optic lines to convey the audio and video signals from point to point.

TELEPORT: A satellite uplink/downlink facility. Access is made available to satellite users, usually on an hourly basis. Gateways are teleports used for international satellite transmissions.

TERMINATION: A 1-volt video signal requires termination by a 75-ohm resistance. Improperly terminated video signals will exceed 1 volt amplitude, and twice-terminated signals will be half of the proper amplitude.

TIME BASE CORRECTOR: A digital video device used to stabilize the output of a VTR so that the signal can be mixed with other sources.

TRANSDUCER: A device that converts energy from one form to another. The microphone, tape head, and loudspeaker are all types of audio transducers.

TRANSPONDER: A channel on a satellite that accepts a video or audio signal from an uplink, amplifies it, and transmits it back to earth. Most communications satellites have 24 transponders.

TRIAX: Short for *triaxial video cable.* Used in situations that require remote video control of cameras at great distances (i.e., multicamera sporting events). Triax allows multiplexed signals to be transmitted great distances over a fairly small diameter cable.

TVRO: Short for *Television receive only.* Dishes that are used to receive but not transmit. Consumer TVRO dishes cost more than $10,000 in 1980, but now they can be purchased for less than $1000.

TWEAKER: Also known as a *greenie*. A small screwdriver (often made of or coated with plastic to prevent shorting delicate circuits) used to make adjustments to video gear. Every good maintenance engineer has a pocketful of tweakers.

U

UNDERSCAN: A feature on some monitors that permits viewing of the entire picture scanned by the CRT. The overscanned part of the picture is normally hidden by the edges of the display. By using underscan, the operator can check the full area of the picture. This is especially useful when framing camera shots or when using the pulse cross feature.

UNIDIRECTIONAL: A microphone pickup pattern that is sensitive to sounds coming from one primary direction.

UPLINK: A sending dish. A transmitter sends its signal to a large parabolic dish antenna that is aimed at the intended relay satellite.

V

VECTORSCOPE: A dedicated oscilloscope used to monitor the color information of a video signal. Hue and saturation are displayed on a circular grid corresponding to the phase and amplitude of the signal's chrominance information. Usually used in tandem with a waveform monitor.

VITC: Short for *Vertical interval time code*. Time code that is recorded within the vertical blanking interval of the video signal. VITC uses a 90-bit word to assign a value to each frame of video.

VIDEO WALLS: A display technology that uses a bank of closely spaced monitors that are individually fed parts of the complete video signal. The viewer sees one large display by looking at the monitor bank.

W

WAVEFORM MONITOR: A dedicated oscilloscope used to monitor and evaluate the video signal. A trace of the video signal is displayed on a scale of –40 to 100 IRE units. Video levels, timing information, and blanking signals can be monitored by using a waveform monitor. The waveform monitor and vectorscope provide an objective reference when setting levels or maintaining video equipment.

WAVELENGTH: A measure of distance from a point on one wave to the same point on the successive wave. In the RF realm, wavelength is determined by frequency and can be computed by using this equation: V (velocity) $= F$ (frequency) $\times W$ (wavelength), where V is equal to the speed of light.

WINDOW: Satellites: The period of time during which a transmission can take place. TBCs: The number of lines of memory that the TBC can store for correcting timing errors.

WINDOW DUBS (BURNED-IN TIME CODE or IN-VISION TIME CODE DUBS): A dub of a tape showing the time code as visible characters superimposed over the video picture. The visible time code must match the time code recorded on the original source footage for the window dub to be of any use.

WORM: Short for *Write once read many*. A type of optical disk recording that was popularized by the success of audio disks (CDs) and video disks (laserdiscs). WORMs, as the name implies, can only be recorded once. This makes them suitable for permanent storage but limits their use in a post-production or creative environment.

WOW: A problem with most any mechanical recording/playback medium, notably turntables or ATRs, that is experienced as a variation in pitch due to speed fluctuations.

WRITING SPEED: The speed at which the record/playback heads contact the magnetic medium. In modern videotape recorders, writing speed is achieved by moving the magnetic tape past heads that are mounted in a rotating head drum assembly.

Y

Y/C: Luminance/chrominance. This designation is used for video signals that keep separate the luminance and chrominance information, thus preventing some of the normal NTSC artifacts, e.g., cross-color and cross-luminance. The S-VHS video format is designed to utilize Y/C video signals.

INDEX

A

PPM. *See* peak program meter
preamp 31–32, 39. *See also* preamplifier
preamplifier 30–31, 33, 60
preroll 132
pressure zone microphone (PZM) 34
prism beam splitter 54
prism block assembly 56
Pro-Digi 167
proc amp 60, 72, 75, 94, 124, 175
programmable read-only memory
 (PROM) 177
progressive scan 158, 196
projection television 157, 195
PROM. *See* programmable read-only memory
proximity effect 37–38
pulse code modulation (PCM) 110,
 121–122, 168
pulse cross 82, 154
pulse dispersion 20, 21
PZM. *See* pressure zone microphone

Q

quadrature subcarrier 64
quadruplex 106, 112–113, 118
Quantel 141, 170, 174–175, 178
quantizing 166–167, 170

R

R-DAT 106. *See also* digital audio tape
radio 2, 6, 8, 10, 17, 179
Radio Advertising Bureau (RAB) 4
radio frequency (RF) 2, 4, 60, 152
RAI 192
RAM. *See* random-access memory
random access 120, 137–138, 144, 167,
 169, 173, 178
random-access memory (RAM) 177
Rank Cintel 142
raster 61
RCA 34, 112
RCA connector 45–46, 66
RCU. *See* remote control unit
read-only memory (ROM) 177
Rebo Productions 192
receiver 6–7, 10, 152, 190–191
Record Industry Association of America
 (RIAA) 168
reference black 75
reference burst 77
reference sync generator 55
registration 58, 92, 175
registration chart 92–93

remote control unit (RCU) 55
remote production 18
resolution 10, 46, 55–58, 60–61, 63,
 117–120, 154, 157–158, 181, 190,
 192–193, 195–196
resolution chart 92
retentivity 106
reverberation 43
rewritable optical disk recorder 120
RF. *See* radio frequency
RGB 54, 56, 60, 63–64, 119–120, 153–155
RIAA. *See* Record Industry Association of
 America
ribbon element 32, 34, 36
robotics 176
ROM. *See* read-only memory
rough cut 134
routing 46
RS-170A 83, 181
RS-232 175, 179
RS-422 175, 179

S

S-DAT 168. *See also* digital audio tape
S-VHS 10, 64, 106, 113, 117–118, 120,
 122, 126, 142
S-video. *See* Y/C
S/N. *See* signal-to-noise
sampling 166, 170, 181
sampling frequency 168
sampling rate 168–169, 181
satellite 2, 12, 14–21, 125, 179, 183, 194
 C-band 14–17
 Ku-band 14–17
 life span 16
satellite master antenna television
 (SMATV) 17
satellite news gathering (SNG) 15, 18
Saticon 57
saturation 54, 63, 79, 124, 156
SC phase indicator 84
SC/H phase 83–84, 143
SCA. *See* subsidiary communications
 authorization
Schreiber, William 196
SCMS. *See* serial copy management system
Scotch magnetic tape 107, 111
SECAM 63–64, 172, 190
second audio program 9
Sel-Sync 110
Sennheiser 38
sensitivity 57–60, 192
sequencing 181